Adventures in

Earth

and

Environmental Science

Practical Manual

Book 2

Dr. Peter T. Scott

First released 2019 all rights reserved Felix Publishing
Felix Publishing 201

Felix Publishing 2019
www.felixpublishing.com.au email:
info@felixpublishing.com

Print copies available from publisher.

ADVENTURES in EARTH and ENVIRONMENTAL SCIENCE PRACTICAL MANUAL Book 2

Also by the author:

ADVENTURES in EARTH and ENVIRONMENTAL SCIENCE Book 1

ADVENTURES in EARTH and ENVIRONMENTAL SCIENCE Book 2

    and companion PRACTICAL MANUAL for Book 1.

ADVENTURES IN EARTH and ENVIRONMENTAL SCIENCE

    (the composite book containing books 1 and 2)

    and accompanying TEACHER GUIDE

ADVENTURES in EARTH SCIENCE

    A traditional Earth Science text incl. astronomy which
    is also available as a series of smaller books:

Exploration Science (Field Geology and Mapping)
Riches from the Earth (Minerals, Mining & Energy) Changing the
Surface (Erosion and Landscapes)
Rocks - Building the Earth
Fossils - Life in the Rocks
A Dangerous Planet (Earth Hazards)
Through Sea and Sky (Oceanography and Meteorology)
Beyond Planet Earth (Astronomy)

2019 Adventures in Earth and Environmental Science Book 2 Practical Manual Digital Edition   ISBN: 978-1-925662-02-3

2019 Adventures in Earth and Environmental Science Book 2 Practical Manual Print Edition   ISBN: 978-1-925662-03-0

Author: Dr. Peter T. Scott

All illustrations, photographs and videos by the author unless stated Cover photo: Andrew Scott of AJS Creative

Registration:

Thorpe-Bowker +61 3 8517 8342 email:
bowkerlink@thorpe.com.au

No part of this publication may be reproduced, stored in a retrieval system, or transmitted in any form or by any means, electronic, mechanical, photocopying, recording or otherwise, without the prior written permission of the publisher.

© All rights reserved Felix Publishing 2019

About the Author

Dr. Peter T. Scott

Dr. Peter Scott is an award-winning of Earth Science of over forty years' experience in both secondary and tertiary education. He holds a bachelor's degree, two master's degrees and a doctorate, including many years on his own research in locating and correlating coal measures. His experience has been as a classroom teacher and Head of Department but he has also worked at the administrative level as Head of Syllabus and as member of State review panels and has served on advisory bodies for several governments and industry bodies. He has also lectured at university in Science teaching. Apart from his exploration during his own field researches, he has also travelled extensively and has visited many places of interest including Antarctica, the Andes, the Amazon, North Africa, the volcanic islands of the Pacific and Asia, Alaska, Hawaii and the American northwest, northern Europe and remote Australia. His studies and interests have also taken him underground into many limestone and lava caves, coal mines and deep metalliferous mines. The Earth is truly a place of adventure – this book encourages the reader to begin its exploration both at home and abroad.

Table of Contents

Introduction   1.
1. Why "Adventures" in Earth and Environmental Science?
2. Studying Earth and Environmental Science;
3. Writing Practical Reports; Safety in the Classroom Laboratory; Safety in the Field.

Chapter 15: Use of Resources and Energy   8.
15.1 Making Charcoal

Chapter 16: Economic Minerals   10.
16.1 Examination of Some Common Metallic Ores
16.2 Geochemistry of Ores and Economic Minerals

Chapter 17: Non-renewable Fuels and Energy   16.
17.1 Examination of Coal
17.2 Examination of Some Petroleum Products
17.3 Demonstration of the Emission of Fossil Fuel Gases
17.4 Nuclear Chain Reaction Data

Chapter 18: Exploration for Resources   23.
18.1 Simulated Aeromagnetic Survey
18.2 Introduction to Aeromagnetic Survey Maps
18.3 Interpretation of Magnetic and Gravity Data Sets
18.4 Interpretation of Radiometric Data Sets
18.5 Interpretation of A Geochemical Data Set for Gold
18.6 Interpretation of Drill Hole Logging

Chapter 19: Mining Economic Minerals   40.
19.1 Virtual Excursion – Open Pit Mining
19.2 Virtual Excursion – Deep Shaft Mining
19.3 Virtual Excursion – Deep Shaft Coal Mining
19.4 Mining – An Historical Perspective

Chapter 20: Processing the Mined Ore   50.
20.1 Separation of a Mixed Ore
20.2 Froth Flotation
20.3 Chemical Extraction of Copper from Copper Ore

Chapter 21: Monitoring and Management   57.
21.1 Monitoring of Local Climate
21.2 Monitoring of Carbon Dioxide Levels
21.3 Monitoring of Air Quality
21.4 Radon Gas Levels
21.5 Water Quality Monitoring
21.6 Total Dissolved Solids – Demonstration
21.7 Turbidity Management

Chapter 22: Renewable Resources   77.
22.1 Sustainability of Tuna as an Ecosystem Resource
22.2 Sustainability of Timber as an Ecosystem Resource
22.3 Sustainability of Surface Water as an Ecosystem Resource
22.4 Case Studies – Surface Water as a Sustainable Resource
22.5 Geothermal Energy as a Sustainable Resource
22.6 Porosity and Permeability

Chapter 23: Renewable Energies   **94.**
23.1 Hydroelectricity and Wind Power
23.2 Solar Power
23.3 Hydrogen Gas as a Fuel Derived from Solar Power
23.4 Manufacture of Ethanol as A Fuel

Chapter 24: The Earth in Motion   **108.**
24.1 Hooke's Law
24.2 Deformation of Rock and the Effects of Temperature
24.3 Compressional Structures

Chapter 25: Volcanoes   **114.**
25.1 The Shape of a Volcano
25.2 Locations of Some Major Volcanoes

Chapter 26: Earthquakes   **120.**
26.1 Earthquake Epicentres and Magnitudes
26.2 Locations of Some Major Earthquakes
26.3 Predicting Earthquakes

Chapter 27: Wind, Rain and Fire   **132.**
27.1 Runoff Coefficient and Flooding

Chapter 28: A Changing Climate   **135.**
28.1 Correlation of Carbon Dioxide Levels and Global Temperature Change

APPENDIX A: Risk Assessment of Practical Work and Excursions   **138.**

APPENDIX B: Excursion Permission Note   **141.**

Introduction - Revised from Book 1

1. Why "Adventures" in Earth and Environmental Science?

The study of the Earth IS an adventure! Studying Earth and Environmental Science involves:

- **Exploration** - whether it be in remote areas or near places of habitation, field work usually involves the breaking of new ground (forgive the pun!). Often in the more remote locations, field work might be over ground which has never been as thoroughly explored as in your study.

- **Exciting Places** - usually outside of the person's usual habitation range and often in very remote places in different parts of the world, in different physical environments such as deep underground (mines and caves), on high mountain ranges and in different climatic conditions (open oceans, hot and cold deserts and jungles).

Photo: Paradise Bay, Antarctica 2011, the remote Argentinian base Almirante Brown

- **Meeting Interesting People** - who are often happy to share their own experiences and culture, not to mention their home, food and local knowledge. One meets a surprising number of people, mostly good people, such as fellow Earth Scientists and students, field workers in the Earth Industries, Government officials; local land-owners and indigenous peoples.

- **Exciting Research** - especially when working on a new project or studying a new location, Earth Science research, as with all scientific studies can be especially motivating (sometimes to the point of happy obsession!). Just like a good detective story, there are things to find out, evidence to gather, step-by-step deductions to be made and finally a conclusion which answers the research question. Looking for gold or gems would be a simple example but it is when the study has a great number of research questions that the field and its companion laboratory work becomes exciting.

- **Studying at many levels** – whether it is a simple prospecting trip or a detailed study of comet movement in the sky over a long time period (NASA leaves this to gifted amateurs!), Earth Science activities can occur at any age and at every level from the school student amateur to the professional Scientist.

Whatever the nature of the study or the level of the student, the Earth is a dynamic and active place with exciting things to see. If Planet Earth is not adventurous enough, there is always the rest of the Universe.

2. Studying Earth and Environmental Science

Earth and Environmental Science cover a wide range of interesting topics, from volcanoes to researching wildlife in Antarctica. The amount of mathematics, often common to Science subjects, has been reduced within the textbook; providing adequate analysis of large amounts of data, but there is still much content to cover in order that a true understanding of the Earth and beyond is obtained.

Because of the amount of content needed to be learned, it is highly recommended that students adopt sound attitudes and practices. These include:

- **Making good, concise summaries of each topic.** Whilst electronic devices such as computers and tablets are excellent for note taking, there is a tendency in modern education for students (and teachers) to upload or cut/paste previously prepared notes. These are good for gathering information as a primary source, but eventually a certain (minimal) form of summary will be required for study purposes and committing a basic set of notes to memory. After all, one must have some information stored in the brain (as well as on Hard Drive) to be able to use when putting into practice what has been learned. The best way is the ancient method of **reading, analysing and extracting the main ideas** and then **writing them down on paper** as a study summary. Doing this on a computer screen has some usefulness but it is not as good for learning as the hand-eye coordination which occurs with physical writing on note paper as a study sheet. Students should organize these sheets so that there is one page per major topic. The use of simple diagrams, charts and lists is an advantage to learning. For long sequences which should be known at all times and not simply retrieved from a data bank on demand, mnemonics are most useful. These may take a simple form of using the first letter of each word in the sequence to make another, simpler word which forms part of a crazy sentence. The crazier, the better. This is then learnt off by rote (try writing it out 100 times!) with the appropriate mental connections made to the real words and sequence.

- **Good study habits** and time management are not natural processes for most students, even those of mature age. A time-management grid or fixed calendar of regular study and assignment preparation could be suggested by the teacher. This would also include the necessary breaks for leisure time and social activities. Such a timetable should be flexible and include such well-known concepts that there should be social/leisure time immediately after classes; study is best done just before sleeping; that a variety of different topics are best done in small amounts rather than a massive amount on the same topic; and that short (written) summaries of the nightly study are best done last. The latter concept does waste a lot of paper but the hand-eye coordination is most useful for memory retention.

- **Completing practical work and assignments on time.** Never leave due work until the last minute. Students should be taught to plan a sequence of preparation e.g. gather notes by library/Internet research and reading, summarizing these notes in a cohesive form, planning the structure of the main work, then writing the assignment in one or several stages so that it is completed well before the due date. Assignments and practical work which occur regularly and often should be started as soon as possible after they have been set. Interruptions due to personal life will occur and if these interfere with the submission of work, extra time should be given by the teacher if they are valid.

- **Noting down ideas**, words etc. not understood should become a regular habit, especially as data recording whilst working is part of the Earth Science method. Difficult questions, new observations and unknown concepts are the source of further research by asking questions. Teachers should encourage their students to always ask questions about what they see and do. There should be a classroom understanding that there is no such thing as either an awkward or silly question. We learn by questioning – even established ideas.

- **Students should work independently and be accountable for their own work.** Although they may have to work and gather data in small groups, students must understand that the final outcome of this work should be their own. It is very easy using electronic media to copy and manipulate other people's original work. Even without computer programs to analyse student work for such plagiarism, it is usually obvious to the trained teacher to see that the level of understanding, terms used and data outlines is not typical of the age nor academic level of the student. Originality, no matter how simple, is the key to a good assignment.

3. Writing Practical Reports

Each school Faculty will have its own view on the recording of any formal practical work completed by the student. It would be hoped that any formal writing, submission and assessment of student practical report will follow a sequence common to that employed later in their studies as post-graduate students or professional Earth Scientists. This sequence is part of the recording phase of the Scientific Method and should be done so that the practical work or experiment can be read and vetted by others and used as a stimulus for further new experimentation and study. A good report or thesis on any research project should follow a logical sequence generate more questions than it answers.

Scientists, like other managers must write down records of their work and present reports. Earth Scientists working for private companies or government agencies write regular reports to their Directors. Research Scientists publish papers (detailed articles for well-known scientific magazines), present seminars and write books about their work. Students write practical reports or theses for schools and universities. Regardless of the type of report, they all follow a similar system: there is an AIM or intent; a detailed account of the METHOD used; some outline of the findings of the research, or their RESULTS; and eventually a CONCLUSION to state what the research proved, how accurate it was and what good it will do.

An example of a sequence of the main components of any written practical reports may be:

1. Giving a **HEADING** and **DATE** (and if required by the teacher, names of co-workers)

2. The **AIM** is written in FUTURE tense e.g. "To ...." There may be one or two aims but never too many. The Aim shows others what you are setting out to show.

3. There may be a need to put in a brief list of **MATERIALS** to show what was used;

4. The **PROCEDURE** is always written in THIRD PERSON (i.e. impersonal e.g. First Person is "I", Second Person is "We" and Third Person does not use personal pronouns at all.) and in PAST TENSE as though the student has already completed the work (because when it is read it **will** have been completed). The Method is best written in point form (some prefer a Flow Diagram but this can be difficult to construct if the method is complex) e.g.

> "1. The specimen was carefully examined with a hand lens;
> 2. Noting its overall shape, an outline was drawn;
> 3. Internal features in the shape were again examined;
> 4. These features were then drawn within the outline;
> 5. Colour and shading was applied; and
> 6. The original specimen's size was compared to that of the sketch to derive a scale in size (e.g. x5)"

5. **OBSERVATIONS & DATA** will include (in third person, past tense) descriptions of ALL observations found using all senses and/or instruments used in detection or any measurement; any data collected; calculations made (some teachers prefer a separate CALCULATIONS section after results); and whenever possible, a DRAWING or SKETCH of any specimen, apparatus or major feature. In Earth Science, drawings of laboratory apparatus are done as in the other sciences such as Chemistry, as 2-dimensional pencil (black ink only if students are good at drawing) using rulers and printed labels in normal ink. Other sketches e.g. of specimen should be done in 3-dimensions in pencil and later coloured pencils and a SCALE of the size (e.g. x1 or x 2 etc.) given. For electronic submissions of work, students are encouraged to use appropriate software for writing the report, doing the artwork and inserting their **own** photographs. If hand-sketched drawings made during the classroom activity are to be included, these should be photographed or scanned **after** completion taking care with colour, size and contrast of the image. Students' own videos may also be included to illustrate important activity but students who have poor computing skills still should be encouraged to submit written reports. The teacher should devise an assessment system in grading reports so that written or electronic submissions are given fair comparisons.

6. **QUESTIONS** if there are specific questions to be answered, it is advisable to also re-write the question and then give its answer.

7. The **CONCLUSION** gives the answer to the aim, any errors encountered and suggestions for improvements. Sometimes, more substantial reports may also have a final section which refers to possible new research.

The first activity (Observation of Pyrite Crystals) is given to you as an example in the way reports will be written in the future. Future activities will have instructions which must be rewritten in the standard manner.

8. Safety in the Classroom Laboratory

Experience has shown that some Secondary and Tertiary students are uncoordinated and socially irresponsible. For this reason, they and rest of the group need to be protected by the application of a set of firm but fair Safety Rules. These should be simple, able to be comprehended by the student and practiced. It is also a good idea to have them posted within the classroom. In time, most groups get into a working pattern and so The Rules become simply standard learning procedures and so a good, working social atmosphere develops. An example of such rules is given on the next page:

SAFETY in the LABORATORY

The following policy is in place to ensure that the laboratories are safe places for students to work in. Respect for self, others and property is first priority. Please ensure that:

1. Safety apparel is to be worn at all times - aprons, good enclosed shoes, eye glasses etc.
2. You do not enter a laboratory without teacher supervision;
3. Unauthorized experiments are strictly forbidden. Variations in procedure must be approved.
4. Food and drink are not consumed in the laboratory;
5. Movement and noise should be kept to a minimum as distractions cause accidents.
6. Items and liquids are not thrown at any time.
7. You do not touch, taste or smell chemicals and minerals only as directed by the teacher.
8. Spillages, breakages and accidents are reported immediately.
9. You know where all safety switches and equipment are located and how they are used.
10. The work area is clean and tidy. When finished, clean apparatus and return with chemicals etc. to appropriate place. Wipe down laboratory

---

In addition, use of electronic equipment such as computers, tablets and cell phones should be appropriate for the learning environment and secondary to the teaching process. Some schools ban the use of cell phones but with a few simple courtesies they can be useful in the classroom.

9. Safety in the Field

This requires more vigilance than in the protected laboratory environment. In the field, away from the classroom, there are other external influences such as:

- the climate, which could be excessive in its heat or cold, windy, wet or subject to sudden change;

- the terrain, which may be steep, rugged, slippery, loose, and full of ravines, caves or mine shafts;

- fast flowing streams or rough seas with unexpected flooding or wave action;

- the vegetation, which may be dense, tangled, thorny and sometimes poisonous; and

- the wildlife, may be dangerous if disturbed or generally a nuisance (such as flies or mosquitos).

Students should also take precautions such as:

- to not wander about and but keep with the group. There is real danger of becoming lost, falling into ravine or mine shaft or other dangers;

- being prepared for the trip, especially about such necessities as:

    CLOTHING - especially adequate footwear, head covering, suitable
        clothing for the climate and sturdy boots;
    WATER and FOOD as appropriate for the trip;
    SPECIAL SAFETY GEAR (e.g. walking poles);
    INSECT REPELLANT and SUN PROTECTION;
    STUDY ITEMS such as Excursion Guide, notepad and pencil.

- forewarn the teacher about any special needs such as allergies and other problems involving outdoor activities; and

- do not indulge in foolish behaviour such as throwing stones or being a distraction.

Dangerous obstacles should be avoided and only safe, secure tracks should be used. Good navigation is essential and the proposed route and time of return should be left with any local authority such as local police or park rangers as well as with the school authorities. A typical set of Field Safety Rules are:

FIELD RULES

When in the field:

1. Listen to all instructions – especially about specific local hazards;

2. Keep with the group – do not wander;

3. Do not enter bodies of water unless told to do so;

4. Do not enter old mines or industrial workings unless told to do so, then with caution;

5. Do not climb cliffs nor stand under or near unstable rocks;

6. Do not throw any objects, especially hammers nor rocks;

7. Watch your step, especially on slopes and in close vegetation;

8. Wear appropriate field clothing at all times. Be prepared for sudden changes in the weather (rain, cold/heat). Carry a waterproof jacket;

9. Watch out for traffic when on or near roads and railway cuttings;

10. Carry own water and some food;

11. Keep movement and noise to a minimum;

12. Do not use cell phones inappropriately. Headphones are not allowed. Take own care of cameras.

Chapter 15: Use of Resources and Energy

EXPERIMENT 15.1   Time: one lesson

MAKING CHARCOAL

<u>AIM:</u> To use laboratory equipment to make a sample of charcoal from wood.

<u>MATERIALS:</u> Bunsen burner, stand and clamp, test-tube, small pieces of wood (wood splints broken are ideal), mat, complete wood splints, tongs.

<u>BACKGROUND:</u>

Prior to large-scale production of coal, much of the metal-working which took place in Europe involved the use of charcoal. This was made by heating timber in a confined space to convert it into a pure, compact form of carbon which could be burned to the high temperatures used in smelting and shaping metals.

<u>PROCEDURE:</u>

1. Break up splints or shavings of wood into very small pieces no longer than about 1 cm.

2. Place them into a large heat-proof test-tube and shake it so that they fall to the bottom and then weigh the test-tube and its contents.

3. Secure the test-tube in a clamp attached to a stand so that the test tube is angled at about 45 degrees as shown below:

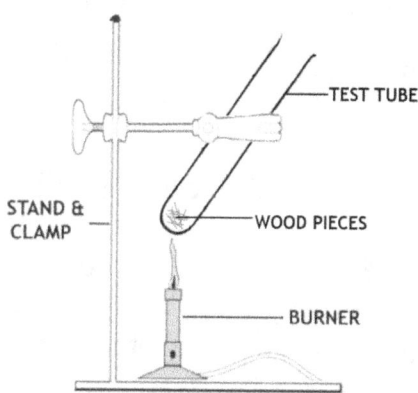

4. Heat the test-tube strongly using the blue flame of the burner.

5. When the sticks start to go black, hold a lighted wooden splint in the mouth of the test-tube.

6. When there is no more reaction shown by the complete blackness of the wood, cool the test-tube then remove it from the clamp and reweigh it.

7. Try igniting the charcoal by holding it with tongs in a Bunsen flame.

EXPERIMENT 15.1 continued

DATA and OBSERVATIONS:

Weight of test-tube and wood before heating: grams

Weight of test-tube and charcoal after heating: grams

Describe the reaction of heating wood in a confined space to get charcoal.

QUESTIONS:

1. What changes were observed to indicate that a chemical reaction has taken place?

2. Was the charcoal lighter than the wood?

3. What happened when the lighted splint was put into the mouth of the test-tube during the reaction? Why?

4. Did the charcoal produced burn when held in a burner flame?

CONCLUSIONS:

1. Write a general conclusion about the use of timber to make charcoal.

2. Apart from making charcoal for later metal smelting and heating, what else was manufactured?

RESEARCH (Optional):

1. Use the Internet to find out how charcoal is made commercially.

2. What is coke in reference to metal smelting? Compare and contrast it to charcoal.

3. Research the uses of (activated) charcoal other than as a source of heat in smelters.

Chapter 16: Economic Minerals

EXPERIMENT 16.1 Time: one or two lessons

EXAMINATION OF SOME COMMON METALLIC ORES

AIM: To examine, describe and sketch some of the main economic ores.

MATERIALS: Set of economic minerals (e.g. haematite, magnetite, pyrite, sphalerite, chalcopyrite, bauxite, malachite/azurite, galena), Mohs' hardness kits, hand lens or binocular microscope, streak plates (unglazed tiles).

BACKGROUND:

Ores are mainly minerals which are pure, inorganic chemical compounds having definite properties. The properties which are useful in identifying ore specimen include:-

1. COLOUR - the frequencies of light coming from the specimen when viewed in normal white light. Simple colours (e g black, red-brown etc) are used to describe this property. A mineral may have a variety of colours.

2. STREAK - the colour of the powdered mineral. If soft, the specimen may be quickly drawn across a streak plate (an unglazed tile -white for coloured minerals, black for white or clear minerals). If the specimen is hard, it may be scratched with a knife or a harder mineral and the colour of the scratch observed.

3. CRYSTAL FORM (habit) – the overall appearance of the specimen in the way that it has been formed e.g. cleavage block (individual crystals not seen, but many flat crystal surfaces observed giving a blocky appearance); botryoidal (many rounded lumps together); fibrous (long strands or fibres); radiating (long crystals radiating outwards from a common point); granular (grouped like grains of sand); amygdaloidal (crystals growing in holes in rock); foliated (as sheets); pisolitic (pea-sized spheres); or massive (no crystals seen).

4. LUSTRE -the way light reflects off the specimen's surface. It may vary from surface to surface (i.e. a specimen may have a range of lustre). Lustre may be:
  metallic -hard, shiny like a metal
  shiny – very reflective like gloss paint
  vitreous -glassy, with some depth
  pearly -like a pearl or button
  silky -shiny with a fibrous look
  greasy -oily appearance like soap
  resinous -with a dull transparent look
  dull -very little shine

5. CLEAVAGE -the ability to break naturally along flat surfaces. The best test is to actually cleave the mineral but this is often not desirable so one has to look for the number of cleavage planes which may meet together. Minerals may have:
  no cleavage (but the mineral may have crystal faces e.g. quartz)
  cleavage in one direction (called basal cleavage which gives sheets;
  two directions giving lines of small steps;
  three directions giving points or corners (which also may be cut off); or in
  four directions giving four-cornered points like a pyramid (rare)

EXPERIMENT 16.1 Continued

6. HARDNESS - resistance of the mineral to be scratched. Usually compared to a standard set of minerals (Mohs' Scale) or a set of material for approximations (Field Scale):

MOHS' SCALE	FIELD SCALE
1. TALC (softest)	Thumb nail 2.5
2. GYPSUM	Coin 3.5
3. CALCITE	Glass 5.0 -6.0
4. FLUORITE	Knife 5.5 -6.0
5. APATITE	File 6.5- 7.0
6. ORTHOCLASE	
7. QUARTZ	
8. TOPAZ	
9. CORUNDUM (ruby)	
10. DIAMOND	

7. SPECIFIC GRAVITY -the density of the mineral compared to that of water, i.e. how many times the mineral is heavier than an equal volume of water (density of water = 1g/cc). Heft is a word to generally describe the mineral's heaviness in vague terms e.g. heavy, medium, light) and can be used in this experiment.

8. OTHER PROPERTIES -any other unique observation e.g. flexible, fluorescent, radioactive, magnetic, chemical reaction, taste (careful!) etc.

SKETCH:

A sketch should be drawn of selected minerals which clearly show some of the above features. These sketches should be to scale and in appropriate colour. This is done by:

a. selecting an appropriate orientation (side) from which the sketch is made;
b. studying the outline and then drawing it lightly to scale so that about two or three sketches will fit per page (you may put a border around each if you wish);
c. re-examination of the specimen (possible with a hand lens) to determine the main internal features for exaggeration and drawing these within the outline. if there is considerable detail, only draw a representative part of it;
d. selecting appropriate colours and then shading the specimen with the pencil on its edge;
e. if the sketch is complete, go over the outline and main features with heavier pencil (or even black ink if you are careful)

PROCEDURE:

1. Carefully examine each specimen and describe its properties in a table using its name and the property names as above i.e. colour, streak, habit, lustre, cleavage, hardness and heft. Also include an additional descriptive column for other if the mineral has any other features worth describing.

2. Draw a sketch in colour of any ore which has a distinct visual appearance e.g. distinct cleavage or habit. Label the sketches.

EXPERIMENT 16.1 continued

DATA and OBSERVATIONS:

1. Describe the ores and their properties in a table.

2. Include several labelled sketches in colour of some specific ores.

QUESTIONS:

1. Why would colour be an unreliable property for identification? Give an example where colour might be confusing?

2. List each ore and state the metal(s) which can be extracted from it.

3. Explain why it may be difficult to describe and identify ore specimens.

4. Are there any specific physical properties other than those main properties listed above which may be useful in quick location and identification of the ore?

CONCLUSIONS:

List each ore and next to it write one or two key words which would be useful in remembering its identification. For example, in the rock-forming mineral quartz previously studied, key words could be glassy and hard.

RESEARCH (Optional):

1. Use the Internet to find out the main mining regions which produce these ores and mark them on a national map.

2. Again, list the metals obtained from these ores and research their uses.

EXPERIMENT 16.2

Time: two lessons

GEOCHEMISTRY of ORES and ECONOMIC MINERALS

AIM: To observe and describe some common tests for metal ions and some salts in common ores.

MATERIALS: Bunsen burner, ceramic mat, paper clips, wooden tongs, test-tubes and racks, powdered minerals (e.g. metal sulfides such as galena and sphalerite; carbonates such as calcite and magnesite; chloride such as halite; and sulfate such as baryte and epsomite), wooden splints, beakers, 2M hydrochloric acid, dropper bottles of barium nitrate solution and silver nitrate solution, de-ionized water.

BACKGROUND:

Chemical testing of various minerals has been on-going for a long time. At the end of the 19th century, German chemists (notably Bunsen and Kirchhoff) were using the newly discovered spectroscope to analyse the light coming from coloured flames produced by burning metal salts in a gas flame. This later led to the development of the Flame Spectrometer which is used today to identify the common metals in minerals. Sometimes the mineral specimens are too small to identify by physical properties, so chemical testing is required.

Several ion groups can also be tested in the school laboratory. Carbonates give odourless carbon dioxide gas in acid and sulfides give smelly hydrogen sulfide (very poisonous! smell VERY cautiously by waving some from the test tube to the nose) with acid. Chlorides give a white precipitate (of silver chloride) with silver nitrate solution (CARE reacts with and stains the skin!) and <u>soluble</u> sulphates give a white precipitate (of barium sulfate) with barium nitrate solution.

PROCEDURE:

Part A. FLAME TESTS (Lesson 1)

1. Set up a Bunsen burner so that it sits on a protective ceramic mat.

2. Unfold a paper clip to a straight wire and, holding one end in wooden tongs, heat the other end in the BLUE flame of the Bunsen burner.

3. Dip the hot end of the wire into water and then dip it into a sample of a metal salt to pick up one or two small crystals (important, use the smallest possible amount).

4. Heat the crystal(s) in the <u>blue</u> flame of the Bunsen burner and note the colour of the flame.

5. Repeat parts 3 and 4 (above) for each of the selection of metal salts. Remember to <u>clean</u> the end of the wire in water after each test and note the colour of the flames.

EXPERIMENT 16.2 continued

Part B. CHEMICAL TESTS (Lesson 2)

1. TEST FOR CARBONATES - using only white calcite and magnesite

1. Into separate test tubes place a small amount (half-pea size or about 4 rice grain size) of each of the carbonates provided. Note the positions and label each sample with their names.

2. Just cover each sample with a little acid (CARE It is caustic and will damage skin and eyes) and observe any reaction (cautiously sniff any gas....there may also be some pungent acid vapour).

3. Hold a lighted wooden splint near the mouth of each test-tube.

2. TEST FOR CHLORIDES - using only halite (Common Salt)

1. Into a test tube place a very small amount (2-3 rice grains) of halite (sodium chloride).

2. Cover with about half a centimeter of <u>deionized</u> water and shake to dissolve some of the mineral.

3. Add two drops of silver nitrate solution (CARE) and observe.

4. Hand in the test-tube for removal and cleaning later.

3. TEST FOR SULFATES - using only baryte and epsomite

1. Into separate test tubes place a very small amounts (2-3 rice grains) of baryte (barium sulfate) and epsomite (magnesium sulfate);

2. Cover each with about half a centimeter of <u>deionized</u> water and shake to dissolve some of the mineral;

3. Add two drops of barium nitrate solution and observe;

4. Hand in the test-tube for removal and cleaning later.

4. TEST FOR SULFIDES

(Students are warned not to make this gas in large amounts and to sniff it very cautiously. IMMEDIATELY after all students in the group have made an observation, the test tube and its contents must be handed in for removal to the fume hood)

1. Into separate test tubes place a very small amounts (2-3 rice grains) of each sulfide.

2. Cover with about half a centimeter of acid and observe. CAUTIOUSLY smell any gas given off.

EXPERIMENT 16.2 continued

RESULTS:

1. Record all of your observations in a table for each part.

2. Find out the chemical name of each mineral tested and the reagents used to test them and write a word equation for each reaction.

3. Draw a representative sketch of about one half to one page in two dimensions with labels of any one of the test-tube reactions.

QUESTIONS:

1. Why were each of the specimens finely powdered?

2. What safety precautions were necessary in this activity?

3. What were the special needs of disposing of the wastes? Why? (some class discussion may be needed here!)

CONCLUSION:

Use the following questions to write a detailed conclusion about this activity:

1. List or make a table showing the specimen name, the test performed on it and its result for each metal ion (flame test) and for each non-metal group.

2. Why was de-ionised water used for making up solutions and dissolving the minerals?

3. What would be the main errors in these tests (be specific for Parts A & B)?

4. In what situations would a scientist in the field use some of the tests in Part B? Explain.

RESEARCH (Optional)

Use the Internet to find out how geochemists analyse sample using flame spectrometers and other devices.

Chapter 17: Non-renewable Fuels and Energy

EXPERIMENT 17.1　　　　　　　　　　　　　　　　　　　　　Time: one lesson

EXAMINATION OF COAL

<u>AIM:</u>　To examine, describe and sketch some of the main fossil fuels.

<u>MATERIALS:</u>　Set of coals e.g. peat, lignite, bituminous coal and anthracite; hand lens or binocular microscope.

<u>BACKGROUND:</u>

Coal is formed in a freshwater, anaerobic environment and come in several ranks depending upon their degree of coalification. During this process of coalification, the amount of ash or silica and water is removed whilst the degree of fixed carbon increases. All ranks have been used as a source of fuel with bituminous coal being preferred for power generation.

<u>PROCEDURE:</u>

1. Carefully examine each specimen and describe its properties in a table using its name; overall colour; feel including powdery nature; internal structure such as fibres, bands etc.; and fracture (even, uneven, conchoidal). Also include an additional descriptive column for other if the mineral has any other features worth describing.

2. Draw a sketch in colour of two or three coals to show their main features and differences. Label any structures seen within the coals.

<u>RESULTS:</u>

1. Record all of your observations as tables for each part.

2. Add a sketch of the coals.

<u>QUESTIONS:</u>

1. In some places, peat and brown coal are used in power stations to generate electricity. What would be the disadvantages of using such fuels?

2. What happens to the ash and water when coal is used as a fuel?

<u>RESEARCH (Optional):</u>

1. Use the Internet to find out about peat: how it was formed, how it is mined and where it is still used as fuel.

2. Use the Internet to research "clean coal technology". How is this done? Are there ways of having a zero emissions coal-fired power station?

EXPERIMENT 17.2

Time: one lesson

EXAMINATION OF SOME PETROLEUM PRODUCTS

AIM: To examine, describe and test some of the main products obtained from crude oil.

MATERIALS: Set of petroleum products e.g. tar or bitumen, heavy oil, light oil, gasoline, kerosene and propane or butane gas, disposable test-tubes with stoppers, watch glasses, ceramic mat.

BACKGROUND:

Coal and oil are both hydrocarbons formed from organic sources. Oil is formed from decomposing marine organisms also in anaerobic conditions. Crude oil is extracted from the ground via deep wells and then refined in fractionating columns in which the various fractions boil off when heated and then condense in cooler areas where they are run off.

PROCEDURE:

1. Examine small samples of each material in separate disposable and stoppered test-tubes (except for the tar which is semi-solid). In a table, write the name of each specimen and describe its colour, clarity by giving a comparative description of how light passes through the liquid; viscosity (slant the test tube and see how fast it runs down the tube...do not allow any to spill). Also add an extra column to record how it burns.

2. DEMONSTRATION: the teacher will place a drop or two of each of the specimen onto a watch glass and ignite it. Students should observe the intensity and colour of the flame and the amount of soot given off. The propane or butane should be from an appropriate device such as a small blow lamp or lighter.

RESULTS:

1. Record all of your observations as tables for each part.

2. Draw a representative sketch of one of the heaviest oils sliding down the test-tube.

3. Draw a representative sketch of one of the heaviest oils sliding down the test-tube

QUESTIONS:

1. What is meant by the viscosity of a liquid?

2. What would happen if the liquid oils in this experiment were all mixed together?

EXPERIMENT 17.2 continued

CONCLUSION:

Comment on the physical properties of the oil products and the corresponding types of flames they produce.

RESEARCH (Optional):

1. What are petroleum products used for in addition to being used as fuels?

2. What alternative renewable products could be used instead of petroleum products from crude oil? What are the alternative fuels which could be used instead of petroleum?

EXPERIMENT 17.3

Time: one lesson

DEMONSTRATION of the EMISSION of FOSSIL FUEL GASES

AIM: To examine the products of hydrocarbon combustion and some problems with gas emissions.

MATERIALS: Apparatus as shown below, source of fuel (e.g. candle wax or an oil from the previous experiment).

BACKGROUND:

Coal and oil are both hydrocarbons and are used as fuels in power stations to generate electricity, in motor vehicles and in factories. The gas emissions as flue gas from coal-fired power stations is considered to be the main source of greenhouse gas emissions and of global warming. Several systems have been developed to remove these gases including deep burial in saline artesian wells and the use of algal ponds to absorb the waste gases.

PROCEDURE:

1. Set up the following apparatus using a Buchner Flask and water stream pump to draw the emissions of the burning fuel through the system:

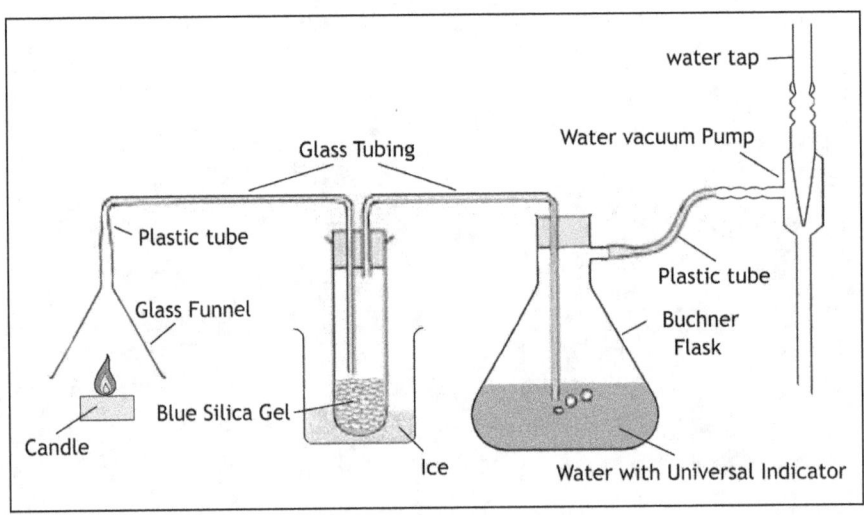

   The water in the flask could be used to represent a deep underground saline solution or a large pond of algae.

2. Ignite the candle or fuel and turn on the water tap to start the suction through the system.

3. Observe all changes in the fuel area, the test-tube and the flask and note these down.

EXPERIMENT 17.3 continued

RESULTS:

1. Record all of your observations.

2. Copy a sketch of the diagram above showing the final changes.

QUESTIONS:

1. What was the purpose of the test-tube in ice and containing the silica gel?

2. Was there any colour change in the Universal Indicator in the flask? What does this show about the gas emitted?

3. Why was the vacuum pump used?

4. If this system was operating using coal or oil as the fuel in a power station, why would the use of such a stage as the ice/silica gel be needed?

CONCLUSION:

1. What gases were given off when the fuel was burned?

2. Write a simple word equation for the combustion of the hydrocarbon fuel.

3. How could this system be made into a closed system?

RESEARCH (Optional):

1. Use the Internet to research the use of closed systems combustion.

2. What is geosequestration? How has it been used in the past? What other ways could be used to totally remove flue gas emissions?

EXPERIMENT 17.4

Time: one lesson

NUCLEAR CHAIN REACTION DATA

AIM: To use measured data to construct a graph of a nuclear chain reaction before it reaches critical mass.

MATERIALS: Data, own graph or graph paper.

BACKGROUND:

A chain reaction refers to a process in which neutrons released in fission produce an additional fission in at least one further nucleus. This nucleus in turn produces neutrons, and the process repeats. The process may be controlled (nuclear power) or uncontrolled (nuclear weapons). In reactors, the rate of the chain reaction is controlled by cadmium and boron control rods which absorb neutrons.

PROCEDURE:

1. Use the data in the following table to plot the number of atoms split against time in microseconds.

TIME	NUMBER of ATOMS SPLIT
1	5
2	15
3	20
4	25
5	45
6	55
7	85
8	120
9	200
10	340
11	800
11.5	1000

2. Construct a line or curve of best fit and hence determine the shape of the graph and the mathematical relationship of number of disintegrations and time.

EXPERIMENT 17.4 continued

RESULTS:

1. Construct a graph similar to the following (do not write on this book):

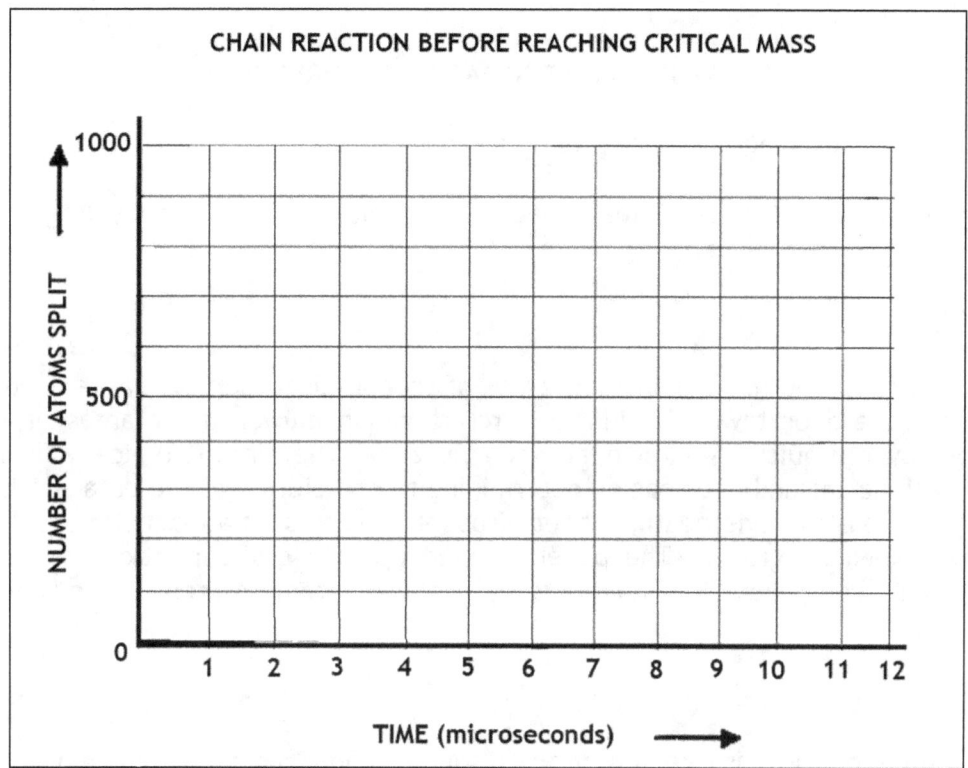

2. Describe the shape of the graph and show the mathematical relationship between time and number of atoms split.

(See http://mathonweb.com/help_ebook/html/functions_4.htm for graphs and relationships)

QUESTIONS:

1. What is meant by critical mass?

2. How is the critical mass attained in a simple fission reactor?

3. How is the chain reaction controlled in a nuclear reactor?

CONCLUSION:

What is the mathematical relationship between time and the number of atoms split?

RESEARCH (Optional)

Use the Internet to locate the closest nuclear reactor. How long has it been in operation?

Chapter 18: Exploration for Resources

EXPERIMENT 18.1 Time: one lesson

SIMULATED AEROMAGNETIC SURVEY

AIM: To simulate an aeromagnetic survey using a simple model.

MATERIALS: A3 sheets of paper, small magnetic compass, small magnet (e.g. refrigerator magnetic strip about 1cm across).

BACKGROUND:

An aeromagnetic survey is a common type of geophysical survey carried out using a magnetometer aboard or towed behind an aircraft which allows larger areas of the Earth's surface to be covered quickly for regional reconnaissance. The aircraft typically flies in a grid-like pattern with height and line spacing determining the resolution of the data. In this exercise a small magnet is pasted underneath a sheet of paper to represent a magnetic ore deposit such as magnetite below the ground. The paper is ruled up in a grid formation to represent the sample positions taken every few moments by the magnetometer represented by the small compass.

PROCEDURE:

1. An A3 sheet of paper is ruled up into grid squares about 5 cm x 5 cm and a small piece of magnetic strip is placed underneath to represent an ore body. Label the grid alphabetically and numerically.

2. Using the small compass, the direction of magnetic north is found on the desk and the sheet of paper lined up so that its long edge points north.

3. Using the small compass as a magnetometer and with it turned so that the north pole on the compass is facing magnetic north, move up and down the paper stopping at each intersection to simulate an aircraft's survey pattern.

4. At each grid intersection, measure the deviation of the compass needed away from its north direction. Measure only the number of degrees deviation from north (000 degrees) without concern for which side the needle swings. Mark this value on the grid intersection.

5. Repeat this exercise until all of the grid intersections have a value. There may be strange readings (anomalies) directly above the "ore body".

6. To increase the accuracy of the plot procedure, extra values can be added halfway between those values on the grid intersections. This is done by taking two values, adding them together then halving them for the value in between e.g. if two grid intersections had values of 56 degrees and 80 degrees then added gives 136 so half would be 69. This value is written exactly halfway between the grid values of 56 and 80.

EXPERIMENT 18.1 continued

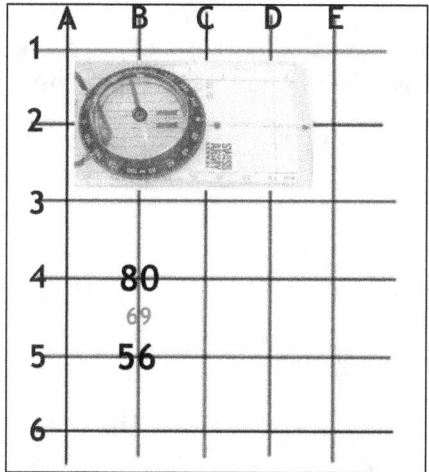

7. Having marked all measured values on the grid intersections and marked all of the averaged values between them, draw curves of best fit around and between these values for a set range of values – say 20, 40, 60, 80, 100 or whatever will fit on the paper depending upon the figures plotted on it e.g. plotting for lines joining points of 70- and 80-degrees deviation:

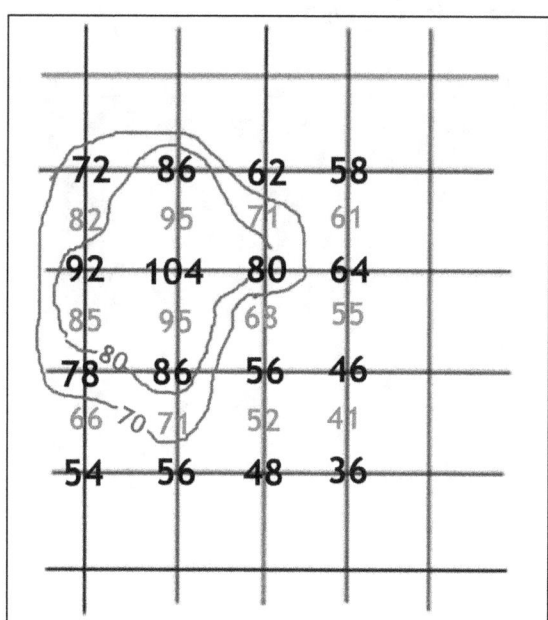

8. Continue to draw these lines which in magnetic survey terms would be lines joining points of equal magnetic strength called isogonal lines.

EXPERIMENT 18.1 continued

RESULTS:

Construct a two-dimensional chart of the pattern of isogonal lines which shows the magnetic anomaly caused by the "ore body" and hence determine the location of it which will be at the place of maximum intensity.

QUESTIONS:

7. What is meant by a magnetic anomaly?

8. What assumption is made by finding the averages between grid values?

9. When the compass is directly over the "ore" the readings may be zero. Why?

CONCLUSION:

What is the map location of the suspected ore body? (Use the alpha-numerical labels e.g. A 1 etc.)

RESEARCH (Optional): Other than searching for ore bodies, what are the other uses of aeromagnetic and ground magnetic surveys?

See: http://www.cas.usf.edu/~cconnor/pot_fields_lectures/Lecture8_magnetics.pdf

EXPERIMENT 18.2

Time: one lesson

INTRODUCTION to AEROMAGNETIC SURVEY MAPS

AIM: To become familiar with some of the properties which indicate sub-surface features on an aeromagnetic survey map.

MATERIALS: Computers and Internet (alternatively have printed colour copies from the two websites listed below)

BACKGROUND:

In geophysics, a magnetic anomaly is a local variation in the Earth's magnetic field resulting from variations in the chemistry or magnetism of the rocks below the surface. Mapping of such variation over a large area is valuable in detecting structures obscured by overlying sediment and vegetation. Magnetic anomalies are usually measured in nanoteslas (nT) or 10^{-9} teslas named in honour of Nikola Tesla. A tesla (T) is an S.I. unit of magnetic field strength or flux, being one weber (wb) per square metre. The weber, named after Wilhelm Weber, is defined as that change in flux per second which will produce an electrical potential difference of one volt. Magnetic anomalies as small as 0.1 nT can be measured in conventional aeromagnetic surveys.

When rock containing magnetisable material, such as iron compounds, is placed in the Earth's magnetic field, the material becomes magnetised and the Earth's magnetic field is reinforced by the magnetic field induced in the rock. The rocks of the Earth's crust are only weakly magnetic and inevitably lie within the influence of the geomagnetic field as it is now, and in some cases, they also record indications of how that field was in the past. Magnetism in rocks can only exist at temperatures below the point where heat destroys magnetism. This is called the Curie point and is variable within rocks of different composition but it is often in the range of 550°C to 600°C, a temperature probably reached by the normal geothermal gradient at depths between 30 and 40 km in the earth, called Curie point isotherm.

Using data from aeromagnetic surveys which record magnetic field strengths as numbers, false colour computer-drawn isogonal maps can be drawn which show various patterns of the magnetic properties of the rocks below the surface. Red colours on the computer-drawn maps represent high positive values of variation from the Earth's field and blues represent low negative values which are less than the average magnetic field of the Earth. In general:

a. Sedimentary rocks have low magnetic values, except when they contain high concentrations of iron oxides from weathering, and so can be useful in finding sedimentary structures such as oil traps in relation to the magnetic fields of basement rocks below them. Banded iron formations and other ore structures rich in iron will show high values of magnetism.

b. Metamorphic rocks are the most common rock found below the surface and they have a large range of magnetic properties usually seen as a complex mix of magnetic intensities, sometimes showing the complex structures within the metamorphic region.

c. Igneous rocks also show a wide range of intensities but granite intrusions (plutons) of uniform composition are often low in value compared to surrounding rocks. Ultramafic rocks high in iron will show high magnetic anomalies and shallow igneous structures such as dykes and sills often show higher values due to their rapid cooling.

EXPERIMENT 18.2 continued

d. Linear features such as dykes and faults are discerned due to their shape and high magnetic colour differences compared to nearby rocks. Faults often have high values due iron-rich serpentine which has come up through the fault from deep below.

PROCEDURE:

1. Closely examine the magnetic anomaly map of part of the west coast of the United States below:

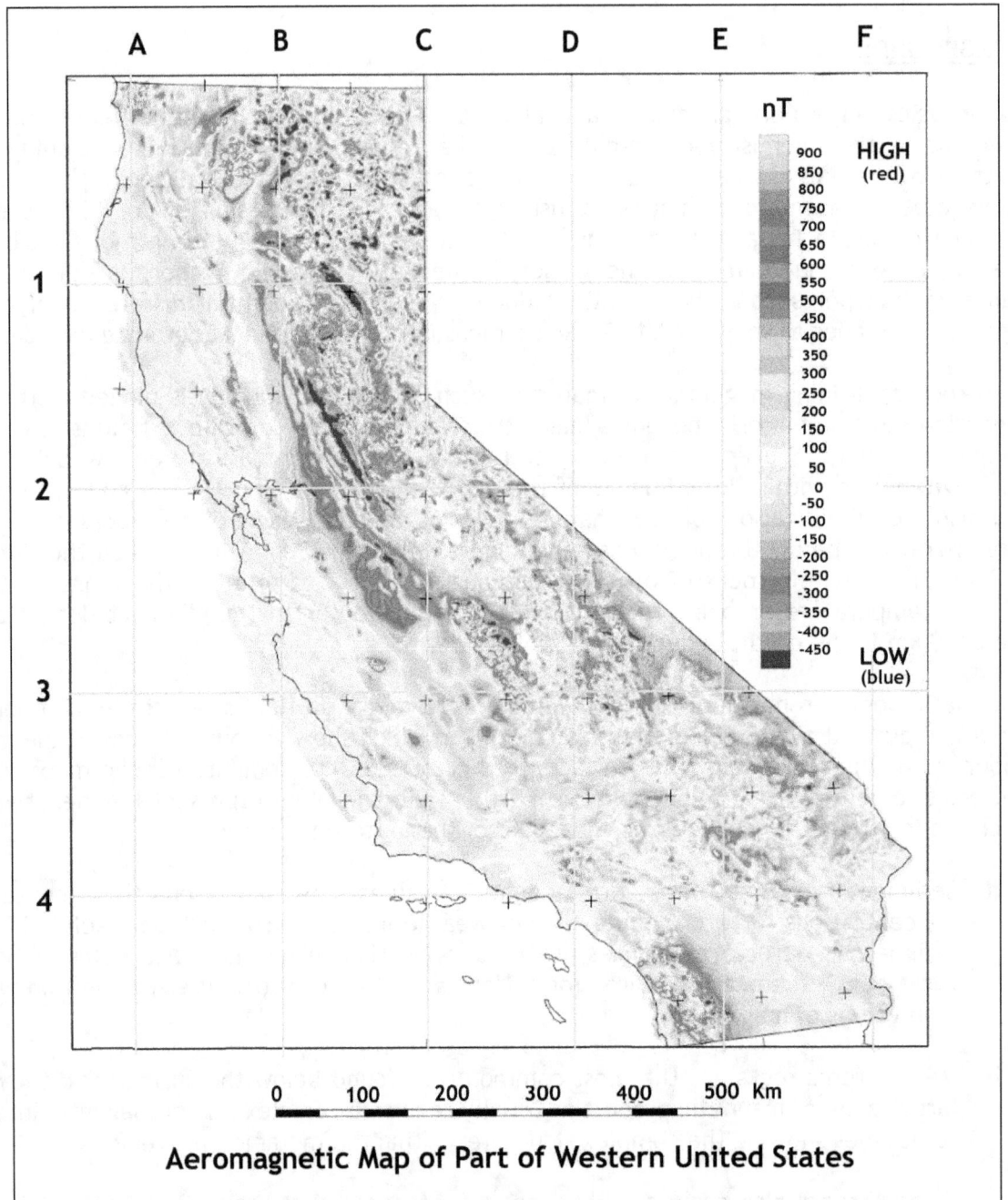

Aeromagnetic Map of Part of Western United States

Data and map from the USGS at https://pubs.usgs.gov/of/1999/0440/pdf/CalMagLg.pdf

EXPERIMENT 18.2 continued

2. Try to find any patterns of high (red) or low(blue) magnetic anomaly e.g. in the top left-hand corner in Square A1 is a rounded patch of red and pink which represents an area of highly magnetic metamorphic rocks. Use the coding given (e.g. A1) to identify any:

   a. Linear features such as faults - list the Square(s) in which they occur.

   b. Other non-linear patches of high magnetic anomaly.

   c. Areas of very low magnetic anomaly.

   d. Other areas of interest.

2. Open the website which shows the surface geology of the area. Comment on the geological features which match the magnetic anomalies identified in part 2 (above).

   http://www.wellsample.com/calif_geol_map.pdf

4. Compare the geological map of the website with the magnetic anomaly map and match any significant geological feature to anomaly areas not yet identified in part 2.

5. Record the matches found in the magnetic anomaly map with the geological map on the website in a table e.g.

MAP GRID ZONE e.g. A1 etc.	DESCRIPTION of ANOMALY high/low/medium, shape	GEOLOGICAL SETTING rocks, age, features	OTHER COMMENTS for identification

RESULTS:

Record all findings in a chart (as above). Also list any other matches found between the two maps. Illustrate any comments or matches with coloured diagrams as required.

QUESTIONS:

1. Magnetic anomaly methodology and interpretation is more complex than what is inferred in this simple exercise. What other factors must be considered when looking at such a map?

2. What are some possible errors which could occur in the making of such a magnetic anomaly map (some extra Internet research may be needed on aeromagnetic surveys?

CONCLUSION:

Write a general conclusion on the use of aeromagnetic surveys and some of the general matches which have been found between the false colours, shapes and the geological setting.

RESEARCH (Optional)

Research the lives and work of Nikola Tesla and Wilhelm Weber

EXPERIMENT 18.3

Time: two lessons

INTERPRETATION of MAGNETIC and GRAVITY DATA SETS

AIM: This is an exercise in examining geophysical data sets with the aim being to provide experience in analysing multiple data sets to interpret regional geology.

MATERIALS: Computer or tablet and Internet connection to the sites listed below.

BACKGROUND:

Exploration geologists usually use a variety of geophysical and geochemical data to locate promising sites for more detailed ground exploration. Magnetic and gravitational surveys from both the air and as ground surveys are just two data sets which can be used. These are often correlated with any known surface geological and topographical (land surface relief) maps to find possible sites for the targeted resource. Aeromagnetic maps have been used in the last two exercises and in this exercise gravitational or Bouguer anomaly maps are introduced. Gravity anomalies are measured using ground or airborne gravimeters and are corrected for the height at which it is measured and the gravitational attraction of terrain such as nearby mountains. This gives a value for the free-air gravity anomaly.

PROCEDURE:

1. Examine each of the following maps carefully. Take some time to read the notes which are provided with each map and make any notes about the methods used and the strengths of the various measurements. The first two maps are geophysical data sets of magnetic anomalies and gravitational anomalies respectively. The third and fourth are topographical (surface features) and geological maps respectively:

    Map A: Magnetic anomaly survey of Australia at:

    https://d28rz98at9flks.cloudfront.net/70282/70282_A0.pdf

    Map B: Gravity anomaly survey map of Australia at:

    https://d28rz98at9flks.cloudfront.net/101104/101104.pdf

    Map C: Surface geology map of Australia at:

    https://d28rz98at9flks.cloudfront.net/73360/Geology_A3.pdf

EXPERIMENT 18.3 continued

2. Compare Maps A and B with the map below showing the main structural elements of Australia and attempt to identify some of the major regions from the magnetic and gravity anomaly maps. Describe how they were identified.

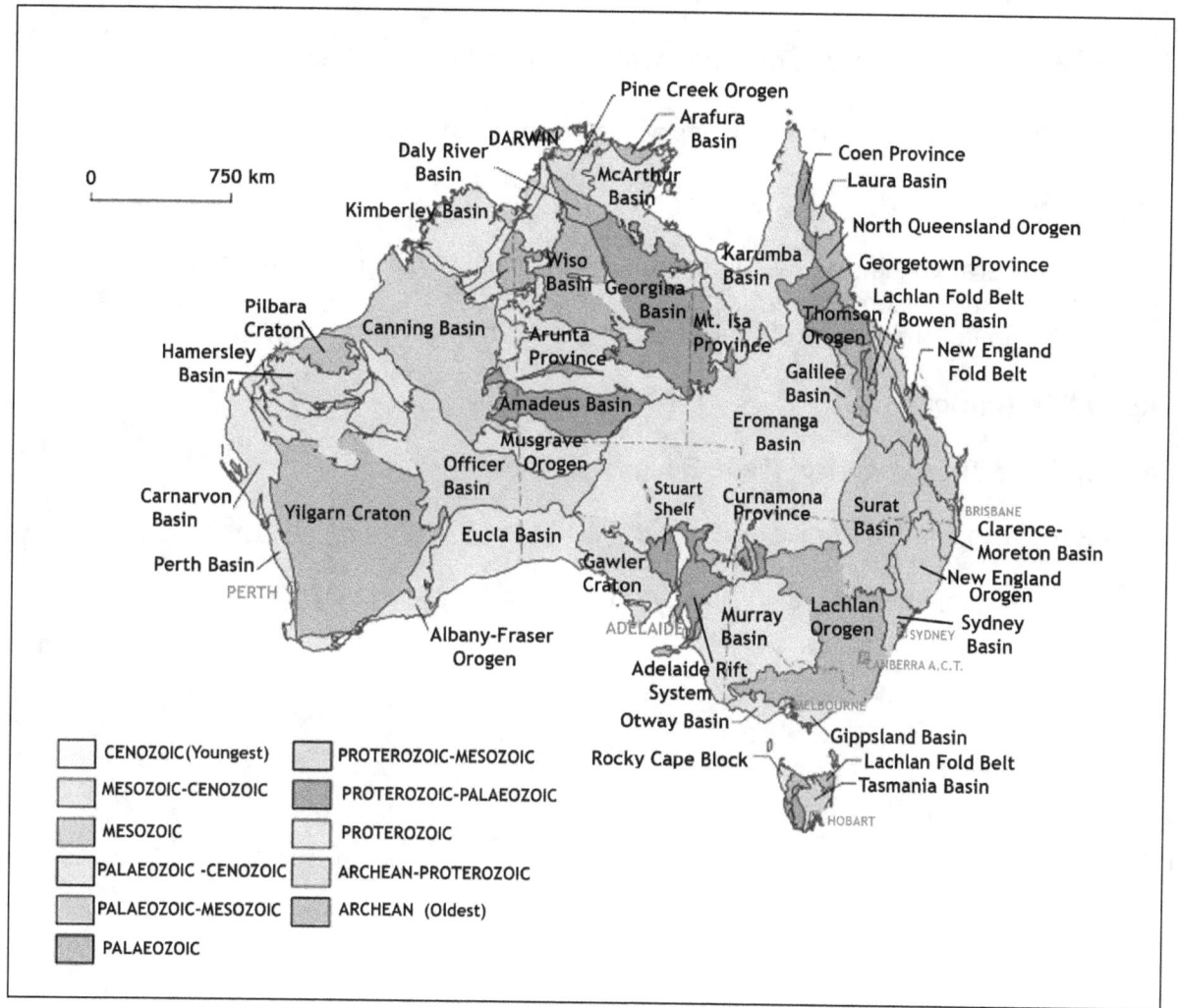

Compare Maps A and B with Map C showing the surface geology of Australia. Is there any match between the anomaly maps (A & B) with any of the major geological settings seen in Map C? Explain.

RESULTS:

Record all findings in a chart similar to that below, but allow room for longer descriptions:

STRUCTURAL ELEMENT (as on map above)	GEOLOGY	MAGNETIC ANOMALY INDICATORS	GRAVITY ANOMALY INDICATORS	COMMENTS

Also list any other matches found between the maps. Illustrate any comments or matches with coloured diagrams as required.

EXPERIMENT 18.3 continued

QUESTIONS:

1. What does Map A show about the island of Tasmania (bottom of map)?

2. Which of the two anomaly maps seemed to show more detail?

3. What units are used in measurements on the gravitational anomaly map?

4. Why are Bouguer anomaly maps adjusted to terrain and height?

CONCLUSION:

Write a general conclusion on the usefulness of geophysical survey maps such as magnetic and gravity anomaly maps referring to some of the general matches which have been found between the false colours, shapes, geological structures and the surface geology.

RESEARCH (Optional)

Research the life and work of Pierre Bouguer

Also see: Explanation of magnetic surveys at: https://youtu.be/AZyNIGFHsE4

Explanation of gravity surveys at: https://youtu.be/9P6GEpxFtSY

EXPERIMENT 18.4 Time: one lesson

INTERPRETATION of RADIOMETRIC DATA SETS

AIM: To provide experience in analysing a radiometric map used in the exploration of uranium and other radioactive metals.

MATERIALS: Computer or tablet and Internet connection to the sites listed below.

BACKGROUND:

The radiometric survey method is another geophysical process used to estimate concentrations of radioactive isotopes of the elements potassium, uranium and thorium by measuring the gamma-rays which these elements emit during radioactive decay. Airborne gamma-ray spectrometric surveys estimate the concentrations of the radioelements at the Earth's surface by measuring the gamma radiation above the ground from low-flying aircraft or helicopters. The data is usually obtained at an elevation of between 40 and 150 metres above ground level and a measurement is recorded every 50 - 60 metres along flight lines. Final data resolution is primarily determined by survey line spacing and elevation. Measurements are made by scintillation counters which register the impact of gamma rays on special chemicals which give off light when struck. This is then magnified using photomultiplier tubes and received as an electrical signal. These can later be processed by computers into false-image maps.

PROCEDURE:

1. Examine Map A, the radiometric map of Australia carefully. Take some time to read the notes which are provided with the map and make any notes about the methods used and the strengths of the various measurements, especially how uranium is detected and the false colour which represents it. Usually red = high concentrations and blue = low concentrations.

    Map A: Radiometric map of Australia at:

    https://d28rz98at9flks.cloudfront.net/70791/70791_5m.pdf

    Map B: Surface geology map of Australia at:

    https://d28rz98at9flks.cloudfront.net/73360/Geology_A3.pdf

2. Look for high concentrations of Uranium on Map A and try to match these locations on the surface geology Map B. On Map A there is also a smaller map at right which is only for uranium. This may be better for comparisons.

EXPERIMENT 18.4 continued

RESULTS: In paragraph form, list the main regions of Australia (e.g. north western New South Wales etc.) and how the radiometrics for uranium matches the surface geology. Indicate if possible, the main types of rocks in the area and their age. Simplify any technical term such as felsic intrusive rocks. Illustrate your answer by any diagram if required.

QUESTIONS:

1. What type of rock often contains high concentrations of uranium?

2. What do the white parts of the map represent? What does this suggest?

3. Australia has some of the biggest uranium reserves in the world and actively exports uranium. Where are the major mines located?
   (See: http://www.world-nuclear.org/information-library/country-profiles/countries-a-f/australia.aspx and scroll down to the map)

CONCLUSION:

Write a general conclusion on the usefulness of radiometric survey maps referring to some of the general matches which have been found between the false colours, shapes, geological structures and the surface geology.

RESEARCH (Optional)

1. What are some of the problems associated with uranium mining in Australia?

2. Prepare a short talk or fact sheet for the case (a) for and (b) against the development of nuclear power in Australia.

EXPERIMENT 18.5   Time: one lesson

INTERPRETATION of a GEOCHEMICAL DATA SET for GOLD

AIM: To provide experience in analysing geochemical maps used in the exploration of gold and other metals.

MATERIALS: Computer or tablet and Internet connection to the sites listed below.

BACKGROUND:

A geochemical survey in mining exploration consists of two particular stages. The first involves collecting and analysing various types of geological materials, such as soils, stream sediments and rocks. The second step, concerns the interpretation of available numerical information by plotting the geochemical values on maps, and the interpretation of the results. After discovery of the mineral deposit, geochemical sampling plays a key role in determining its extent and potential. Sometimes mineralization may not be obvious and impossible to recognize in hand specimen. Without the use of geochemical sampling methods, many known ore deposits would probably not have been discovered. The collected materials may be analysed for any number of elements and further action depends upon the demand for a particular resource, the geology of the area, and the potential for exploitation.

PROCEDURE:

1. Open the site of the National Geochemical Survey of Australia at:

    https://d28rz98at9flks.cloudfront.net/71973/Rec2011_020_Vol1.pdf

2. Explore this site noting the sampling techniques (scroll to page 5), the collection and preparation of the samples (page 7) and their analysis (page 8). Make a short summary of this information as an introduction in the Conclusion.

3. Scroll down to page 10 and in the Results copy out the properties listed for gold (Au).

4. Scroll down to the map showing gold sampling distribution and values on page 22. Examine the distribution over the continent of the major concentration level i.e. 0.087 mg/kg. Comment on the locations of these concentrations. To assist in this section, go to: http://www.australianminesatlas.gov.au/aimr/commodity/gold.html and scroll down to the map. Relate the geochemical analysis to current gold mining areas and any other locations which may have potential.

5. Continue exploring this last website with special reference to the trends in the production and value of gold in Australia and the local State on the world market. Scroll down to Table 1. and graph the gold production in tonnes for the years 2006 to 2012 for local State (e.g. QLD, NSW etc.).

EXPERIMENT 18.5 continued

RESULTS:

List the main properties of gold.

Comment on the main locations of high values of gold in Australia as determined by the national geochemical survey and how they relate to current gold mining locations.

Draw a simple line graph of recent gold production for the local State (e.g. QLD, NSW etc.)

QUESTIONS:

1. How were the samples collected? (refer to number of sample sites and the meanings of TOS and BOS).

2. How were the samples prepared and analysed? What precautions were taken to prevent errors?

3. Comparing gold to other metals, what is a unique property which would aid in its sampling?

4. Is there a match between gold mining locations and this new sampling data set? Give some locations which have a good match.

5. Are there any locations on the National Geochemical Survey map which are not currently mined? Comment on their potential.

CONCLUSION:

1. Write a general summary of the methods used in the national survey.

2. Write a general conclusion on the usefulness of geochemical survey maps referring to some of the general matches which have been found between the survey and known gold mining locations.

RESEARCH (Optional)

Use the Internet to find out about how gold is used.

EXPERIMENT 18.6

Time: one lesson

INTERPRETATION of DRILL HOLE LOGGING

AIM: To provide experience in interpreting drill core electrical logs.

MATERIALS: Paper, pens, rulers.

BACKGROUND:

The data set given in this exercise is a simple representation of an electric log. It consists of a Spontaneous Potential (SP) curve and two Resistivity curves of varying depths of investigation; the short spacing and long spacing resistivity curves. E-logs are practical, low-cost logging tool useful in supplementing drill core logging as well as providing data in the exploration of metallic ores, oil, coal and water.

The Spontaneous potential (SP) log measures changes in electrical potential (voltage) between an electrode on the sonde lowered down the well and one at the surface. An electrical potential exists between the natural pore fluids of the rock and the drilling mud, where these come into contact through the semi-permeable barrier of porous rock. Therefore, SP logs are a measure of rock permeability with:

- Shales and limestone are nearly impermeable and have a positive reading with the curve moving to the <u>right</u>. Fluids with high resistivity (i.e. lower conductivity) such as freshwater also have a positive reading, often showing a higher positive reading than shales.

- Negative deflection occurs for rocks which have a high permeability, such as sandstones (sands) or fluids with higher conductivity than the drilling mud such as saltwater and oil.

Short spacing and long spacing Resistivity Curves are typically used in electric logging along with SP logs. Resistivity is the reverse of electrical conductivity which depends upon fluids in the rock pores and how they conduct electricity. The depth of penetration is related to the distance between the electrical current electrode and the other potential electrode. The short spacing or 16-inch (41 cm) resistivity curve may be influenced by invasion of the drilling fluid into the formation but long-normal or 64-inch (163 cm) resistivity curve shows the resistivity of the same zone further away from the borehole, beyond the invaded zone. In permeable soil or rock, the porosity affects the depth of invasion and the dissolved ion content of water in the pores determines the resistivity of the material. The two curves are plotted side by side for comparison. For Resistivity:

- Curves with low resistivity values to the left are due to shales or sands saturated with highly-conducting saltwater which will give lower values than shale.

- Curves with high resistivity values move out to the right are usually due to sands (sandstones) and even higher indicate compact limestones. Freshwater in sands gives a high value as does oil but this is less than freshwater.

EXPERIMENT 18.5 continued

PROCEDURE:

1. Carefully examine the simplified E-log below:

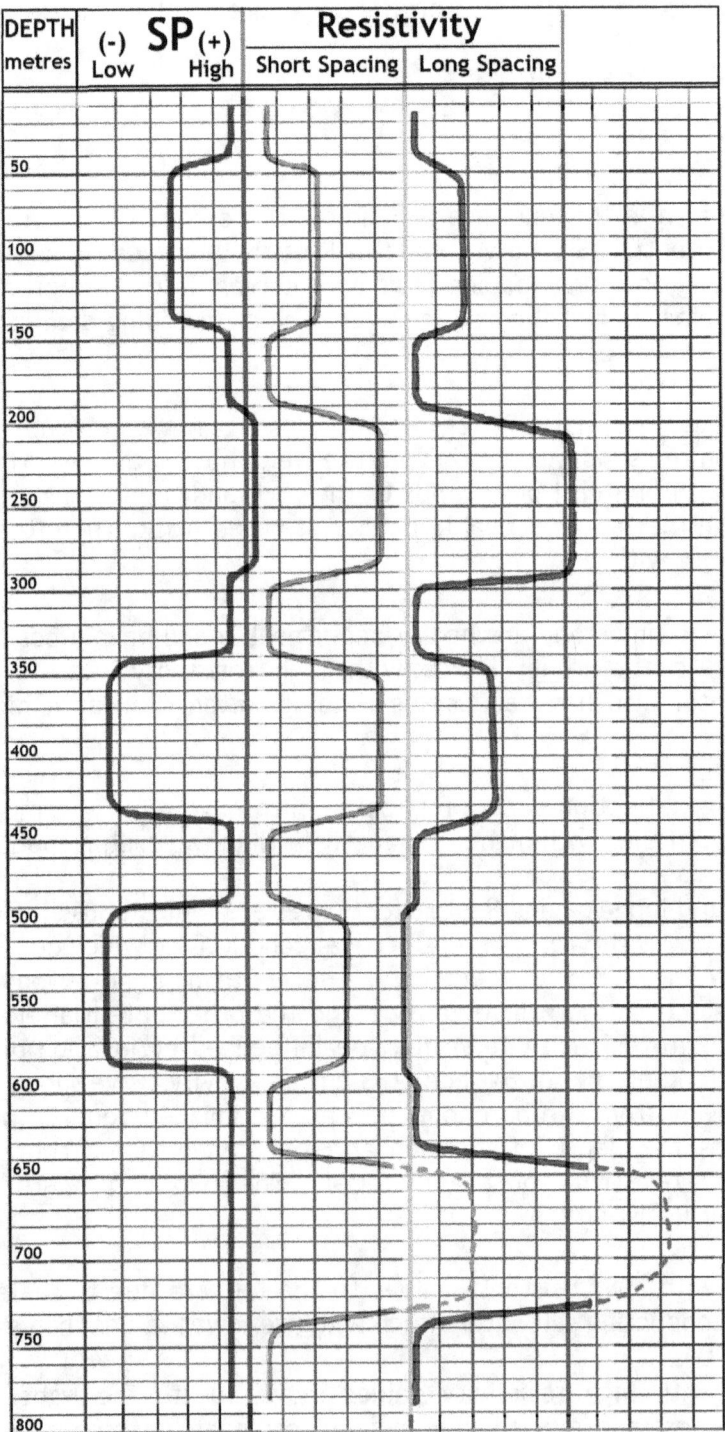

EXPERIMENT 18.5 continued

2. Knowing that the region was once a marine environment, use the SP and Resistivity curves to construct a generalised stratigraphic column by drawing another column to scale (say 1.5 cm = 50 metres depth) for the same depth as the log. As shales are often impermeable and not influenced by any fluid within them, they are often consistent in their position and a Shale Line is sometimes used as a guide which is drawn down the length of the log at the consistent position of the shales. The lithology can be determined from the curves remembering that:

Lithology	SP Curve	Resistivity Curve (long spacing)
Shales	Consistently strong positive (right)	Low (left)
Sandstones	Low due to their good permeability and may contain oil or water.	Higher resistivity especially with freshwater and slightly less so for oil. Saltwater in sands gives a very low values, less than the Shale Line.
Shaly sands	Intermediate for dry shales and sands.	
Compact limestones	Same as for shales - along the Shale Line	Extremely high. Higher than sands.

3. Knowing the lithology, locate the depth at which a drill will first encounter (a) an aquifer of freshwater and (b) an oil trap with oil.

RESULTS:

1. Draw the stratigraphic column to scale. Use conventional symbols and colours to represent the lithologies and show these as a legend.

2. Give the depth to the (a) freshwater aquifer and (b) the oil trap.

QUESTIONS:

1. What is the relationship between electrical resistivity and conductivity? What is the unit of resistivity?

2. What factors determine the resistivity of a rock or sediment?

3. Why would crude oil have lower resistivity than freshwater? (some research needed)

4. Why would the long spacing resistivity results be more reliable than the short spacing?

5. Why is the curve for shale often used as a standard Shale Line for comparisons?

EXPERIMENT 18.5 continued

CONCLUSION:

1. Write a general summary of the usefulness of e-logs in exploration geology.

2. Write a specific evaluation (as an exploration geologist) of the geological setting and its potential for exploitation, giving details of the lithology, any resources found and any other factors about their extraction.

3. What are some of the errors which may give inaccurate results in using this method? How could they be compensated?

RESEARCH (Optional)

Use the Internet to find out about other types of electric logs which could be used, especially those which may reinforce the findings of this SP/Resistivity log.

EXPERIMENT 18.5 continued

Chapter 19: Mining Economic Minerals

EXPERIMENT 19.1 Time: one or two lessons

VIRTUAL EXCURSION - OPEN PIT MINING

AIM: To use online videos to find out about the operation of a typical Open Pit or Open Cut mine.

MATERIALS: Computers or Tablets, Internet connection.

BACKGROUND:

Open pit mining is used when the resource is relatively close to the surface, when the overburden or waste material above is thin or when the resource is wide-spread and unsuitable for tunnelling. The rock is primed with explosive charges and broken up, then conventional but large-sized plant machinery such as excavators, bulldozers, front-end loaders and dump trucks are then used to carry the resource and gangue to the separation plant where the resource is removed and the waste material returned to the pit for back-filling. Bulldozers and graders are then used to reshape the land surface which is then covered with the topsoil which had been removed and stockpiled before the mine began operations. Native vegetation from seeds collected by the environmental team before the start of the mine, is then replanted.

PROCEDURE:

Part A: Cadia Gold and Copper Mine, New South Wales, Australia.
This video was made by the author during a student visit sponsored by the Australian Institute of Mining and Metallurgy. Open Pit mines usually cover large areas either because the resource such as coal is in extensive flat, nearly horizontal seam or because the metallic resource is spread throughout the country rock.

1. Open the site of the video at:

    https://www.youtube.com/watch?v=ohxxYSjcXa4

2. Watch this video carefully and note how the material is extracted, transported and stockpiled. Note any other details of interest about the operation, especially safety and environmental concerns.

3. In the Results summarize the operation using the captions shown on the video.

Part B: Gold Mine, Nevada, USA - an extensive virtual excursion.
Sometimes the resource is mined to the limit of Open Pit mining and so an underground mine is started at the bottom of the pit. This excursion visits a much larger mine and offers a much more detail look at the operation. Take the time and opportunity to explore this site fully with special reference to the life and work of the people engaged in the operation.

Open the site at:

   http://clickschooling.com/2017/10/nevada-gold-mine-virtual-tour/

EXPERIMENT 19.1 continued

1. Read the introduction page carefully before continuing on at the link at:

    http://xplorit.com/nevada-mining-web#

2. Explore the website, looking for details of the method of operation, the machinery used and how the people work in the mine. A detailed list of observations can be made in the Results.

Part C: Black Thunder Open Pit Coal Mine – is in the U.S. state of Wyoming, located in the Powder River Basin and contains one of the largest deposits of coal in the world. In large Open Pit coal mines, much of the same operation occurs but the coal is usually sieved on large screens and washed before the final storage phase.

1. Open the site of the video at:

    http://www.rmcmi.org/news/detail/2011/05/03/take-a-virtual-tour-of-black-thunder-coal-mine#.XC2T4WlLfX4

2. As before, watch this video carefully and note how the material is extracted, transported and stockpiled.

3. In the Results describe the operation using the captions shown on the video.

RESULTS:

Comment on any useful information obtained from the mines at locations in each of Part A, B and C.

QUESTIONS:

1. In general, and from the textbook notes and the videos, what would be the sequence of events before the mining company arrives on its site?

2. What happens next after the environmental phase when seeds and seedlings are collected from the area and escape corridors are made to allow animals to leave the area?

3. Having exposed the rock surface, how is the mining process started?

4. Why are the haulage roads made in a spiral manner around the outside of the pit? What would be some of the safety considerations and strategies used to ensure the safety of truck drivers and the stability and surface of the roads?

5. How is the material exacted and transported? Where is it taken?

EXPERIMENT 19.1 continued

CONCLUSION:

Summarize the main operations observed in the Open Cut process. A Table giving the various phases of operation for each of the three locations may be useful.

RESEARCH (Optional)

Use the Internet to find out about the locations of each of these mines and the minerals mined at Parts A and B.

EXPERIMENT 19.2   Time: one or two lessons

VIRTUAL EXCURSION - DEEP SHAFT MINING

AIM: To use online videos to find out about the operation of a typical Deep, Shaft mine.

MATERIALS: Computers or Tablets, Internet connection.

BACKGROUND:

Deep Shaft mining is a form of underground mining using shafts driven vertically from the top down into the earth to access ore or minerals. Shaft mining is particularly ideal for concentrated mineral deposits, especially those which are in non-uniform shape such as gold within quartz veins. They are also used if the ore deposit or coal seam is near or below valuable pastoral land or areas of urbanisation. A shaft mine is usually comprised of several separate access points including a primary access or central shaft with a cage which functions as an elevator to transport equipment and workers in and out of the mine and a secondary access called a skip to transport the mined ore or other material up to the surface. A third shaft is reserved as an emergency exit, housing an additional cage or system of ladders. Supplementary shafts or compartments in the mine service cables and pipes for transporting water, compressed air, and diesel fuel. Additional horizontal workings referred to as drifts, galleries, or levels extend out from the central shaft in the direction of the ore body.

PROCEDURE:

Part A: Central Deborah Mine Gold Mine, Bendigo, Victoria, Australia.
This video was made by the author during a student visit to this deep gold mine, the upper levels of which are often open to visitors.

1. Open the site of the video at:

    https://www.youtube.com/watch?v=k240TTg9F60

2. Watch this video carefully and note how the levels are excavated using explosives and drilling equipment. Also note how the roof of the tunnels are supported and the methods of removing the ore from the mine. Note any other details of interest about the operation, especially safety equipment and procedures.

3. In the Results summarize the operation using the captions shown on the video.

EXPERIMENT 19.2 continued

Part B: Mponeng gold mine in South Africa's Gauteng province - the world's deepest mine. This is a commercial documentary lasting about 30 minutes.

1. Open the site of the video at:

    https://www.youtube.com/watch?v=6ZtYInuOKtE

2. As before, watch this video carefully and note how the shafts and tunnels are excavated and the material is extracted and transported.

3. In the Results describe the operation briefly, especially the problems associated with very deep mines of a large size and the reasons for such deep mines.

RESULTS:

Comment on any useful information obtained from the mines at locations in each of Part A and B.

QUESTIONS:

1. In what circumstances are deep shaft mines used?

2. What type of equipment is used in deep shaft mining? Compare this type of mining with that of Open Pit.

3. What are some of the major problems associated with deep shaft mining concerning:

    (a) construction
    (b) prevention of collapse
    (c) working conditions
    (d) safety

4. Why are deep shaft mines hot? Research how they are cooled to help working conditions.

5. How is the material exacted and transported?

CONCLUSION:

Summarize the main operations observed in the Deep Shaft mining process. In the summary include the reasons for such mining, the methods used in excavating the mine, any engineering processes designed to hold the mine together, how material is excavated and removed and the main problems associated with such mining.

RESEARCH (Optional)

Use the Internet to find out about the locations of deep shaft mines in the local area, what resource they extract and how this resource is transported and marketed. Also research the dangers of abandoned mine shafts in the local area or in general.

EXPERIMENT 19.3   Time: one lesson

VIRTUAL EXCURSION - DEEP SHAFT COAL MINING

AIM: To use online videos to find out about the different techniques used in the operation of underground coal mines.

MATERIALS: Computers or Tablets, Internet connection.

BACKGROUND:

Coal can be mined from surface Open Cut pits or from deep, underground mines. Coal is mined underground when it is too deep for Open Pit operations or the land above the seams is too valuable to excavate in large pits or the seams are below urban areas, rivers and lakes and even below the sea in some coastal mines. Unlike gold and other metallic ore mines, subsurface coal mines are usually more regular in shape as the coal is often found in several horizontal seams. This allows for two major types of excavation operation: Bord-and-Pillar, also called Room-and-Pillar; and Longwall Mining. Bord-and-Pillar excavation involves the removal of coal from tunnels (Bords or Rooms) leaving sections as pillars to support the roof. Longwall operations use an advanced system of hydraulic roof supports and an excavating tool mounted on a long track. This fully-mechanised technique is used when the seams are thick and uniformly horizontal.

PROCEDURE:

Part A: Bord-and Pillar Mining - Great Northern Seam, Lake Macquarie, central New South Wales, Australia.

1. Open the site of the video at:

    https://www.youtube.com/watch?v=od87_PVwPP4

2. Watch this video carefully and note how the mine is entered, how the levels are excavated using excavators, shuttle cars and conveyor belts. Also note how the roof of the tunnels are supported and what happens after the coal is removed from the mine. Note any other details of interest about the operation, especially safety equipment and procedures.

Part B: Longwall Mining, United States.
There is no commentary on this video so notice how the system works, especially the excavating cutter and the use of the hydraulic supports above.

3. Open the site of the video at:

    https://www.youtube.com/watch?v=WmwEB4DY_jc

4. Watch the video carefully and notice how the coal is cut and removed by conveyor. Also notice how the roof is supported and the function of the miners.

5. In the Results summarize both operation by comparing and contrasting the processes, including the equipment used, how the coal is removed, the roof supported and the function of the miners.

EXPERIMENT 19.3 continued

RESULTS:

Summarize the processes used in Bord-and-Pillar and Longwall coal mining. Compare and contrast both processes. Use diagrams as needed to make the summary.

QUESTIONS:

1. What circumstances would determine which type of coal excavation is used?

2. What type of equipment is used in each method? Compare underground coal mining with that of surface Open Pit coal mining.

3. What are some of the major problems associated with Bord-and-Pillar mining concerning:

    a) construction
    b) prevention of collapse
    c) working conditions
    d) monitoring of the mine environment and safety

4. How is the material exacted and transported from the mine?

CONCLUSION:

Summarize the main operations observed in the underground coal mining process. In the summary include the reasons for such mining, the methods used in excavating the mine, any engineering processes designed to hold the mine together, how material is excavated and removed and the main problems associated with such mining.

RESEARCH (Optional)

Use the Internet to find out about the locations of coal mines in the local area, State and Nation. How is this coal transported and marketed and what is its use?

EXPERIMENT 19.4 Time: one lesson

MINING – AN HISTORICAL PERSPECTIVE

AIM: To use online videos to find out about the life and times of mining communities and how they have led to the development of nations.

MATERIALS: Computers or Tablets, Internet connection.

BACKGROUND:

Many countries have a strong history of mining which has given those countries a rich heritage. In the 19th century, major gold rushes took place in Australia, New Zealand, Brazil, Canada, South Africa and the United States. In these countries, people came from all over the world to make their fortune. Most did not and some returned to their homelands whilst others stayed and helped develop their new home. This was the case in the Californian Gold Rush of 1848. In 1851, Edward Hargraves returned to the British colony of New South Wales from the American gold fields and discovered gold at Ophir west of Sydney. Soon the Australian Gold Rush began with large numbers of free settlers arrived in this British convict colony from Britain, Europe, America and China to seek their fortune. Gold was discovered in large amounts at Bendigo and Ballarat in Victoria further south and late far to the west in Kalgoorlie, Western Australia.

PROCEDURE:

Part A: Go back in time to the 1851 gold-mining town of Sovereign Hill, Ballarat in the (then) colony of Victoria.

1. Open the site of the video at:

    https://www.youtube.com/watch?v=xLQUb9tarNA

2. Watch this video carefully and note the buildings, colonial dress, buildings and streets and the general lifestyle of this 1850's gold town.

3. Make brief notes in the Results of the mining operations, the equipment used and how the gold was extracted and processed.

RESULTS:

1. Make comment on the lifestyle of the colonial town in paragraph form. Use small diagrams or sketches if needed.

2. Prepare notes so that a flow chart can be constructed in the Conclusion to show how the gold was processed from the lumps of gold-bearing quartz extracted from the mine to the final process of producing a gold bar. (some research may be needed to understand the use of mercury)

EXPERIMENT 19.4 continued

QUESTIONS:

1. Where is Sovereign Hill located and why did it develop there?

2. What was the significance of the 1851 Gold Rush to the development of Australia?

3. What were some of the differences observed between the lifestyle of the 1850's and today? List all the differences seen or inferred.

4. What type of equipment was used in obtaining gold from (a) the surface and (b) underground?

CONCLUSION:

Summarize the main features of the colonial town of Sovereign Hill observed in the video and suggest how different conditions were then compared to today. Outline the process of refining the gold using a flow chart.

Research (optional)

1. Use the Internet to find out about the Victorian Gold Rush.

2. What contribution did gold rushes make to the countries where they occurred. Discuss.

Part B: Additional Activity: Life on the Victorian Goldfields Game.

AIM: To experience some of the problems of living on a goldfield in the 1850's through gaming.

BACKGROUND:

Living on the gold fields could be hard and sometimes dangerous. There were many people from all over the world crowded into a small area all trying to get rich. The miners came from many countries which had different languages and customs and so there was some conflict between miners and also with the authorities who imposed a system of licences to control mining. This banded the miners together regardless of their personal backgrounds. In 1854, the miners rebelled against the British authorities because of the brutality of the local Troopers enforcing the licencing system and their lack of representation in local affairs. They built a defensive stockade on a hilltop near Ballarat, Victoria and burnt their licences. Under the leadership of Peter Lalor, they raised the Southern Cross flag and were ready to defend their freedom. The miners were further supported by the arrival of two hundred American miners who called themselves the Independent Californian Rangers, under the leadership of James McGill. At 3 am on Sunday, 3 December, a party of 276 soldiers and police, under the command of Captain John W. Thomas attacked the Eureka Stockade and a battle ensued. The ramshackle army of miners was hopelessly outclassed by a military regiment and was routed in about 10 minutes with 22 miners and 6 police and soldiers being killed. There were many casualties on both sides and 114 surviving miners were arrested. At their trial for Treason, public opinion and justice saw the miners acquitted and released. Peter Lalor, their leader was elected as a Member of the Colonial Parliament for Ballarat and the Eureka Stockade, its flag and ideals became part of the Australian psyche.

EXPERIMENT 19.4 continued

PROCEDURE:

Open up the site for the Gold Rush game at:

http://www.nma.gov.au/av/goldrush/

Play the game individually at home or as a group in class. Try several characters to see their point of view. Note some of the problems and history of living in such times.

QUESTIONS and RESEARCH:

There are many questions which will arise from playing the game. Notes these down and then discuss them at the end of the game.

CONCLUSION:

Make a general comment about the game and how it demonstrated how people lived on the goldfields.

RESEARCH (Optional)

Use the Internet to discover the life and times of miners in (a) the California Gold Rush and/or (b) the Australian Gold Rush.

Chapter 20: Processing the Mined Ore

EXPERIMENT 20.1 Time: one lesson

SEPARATION of a MIXED ORE

<u>AIM:</u> To use a sequence of physical separation techniques to extract valuable resources from a mixture of ore and gangue.

<u>MATERIALS:</u> There is a variety of materials to use as a free choice. Some items may not be needed. Materials include the mined 'ore' (a crushed mixture of galena or lead shot, salt, iron filings and sand) and the following items:

- large beaker (500 ml) and small beaker (250 ml)
- gold pan (or small dish with sloping rim)
- plastic tote box or other large rectangular waterproof box with sides
- strong bar magnet
- plastic bags
- Bunsen burner and tripod
- filter funnel, filter paper and conical flask
- source of water e.g. lab. tap
- large watch glasses or Petrie dishes
- accurate electronic balance

<u>BACKGROUND:</u>

Mining occurs because there are some materials which are considered to be a valuable resource. Unfortunately, these resources often come mixed with other resources and with unwanted material called gangue. After an ore is mined, it is then crushed into smaller pieces and then processed in such a way that the basic resources for later refinement is separated from the gangue. Often, the separation techniques used rely on differences in the physical properties of the resources and the gangue material.

<u>PROCEDURE:</u>

This is a do-it-yourself activity! Look at the mixture and the resources which are needed to be extracted i.e. galena (lead sulfide ore), iron and salt (sodium chloride) from a known amount of mixture (e.g. say about 200 ml in a beaker) which is weighed before the separation process begins. Discuss with the group the physical properties of EACH resource which makes it different from the others and the gangue material (sand). Think of the most efficient sequence of events which could be used to extract each of these useful resources from the gangue.

Write up the Procedure in the usual way in point form, third person and past tense. Some assistance may be needed in the techniques of panning and filtration (not in order):

- **Panning.** Watch the technique demonstrated in the author's video on panning at:

    https://www.youtube.com/watch?v=B5h7H4nG7aM

EXPERIMENT 20.1 continued

In this exercise the panning will be done with MINIMAL water into the tote box because the water also needs to be collected (why?)

- **Filtration** uses the standard laboratory process involving a filter funnel with filter paper and a filter flask to receive the water. If too much water and gangue has been collected, then several, separate filtrations will have to be made.

How to fold a filter paper

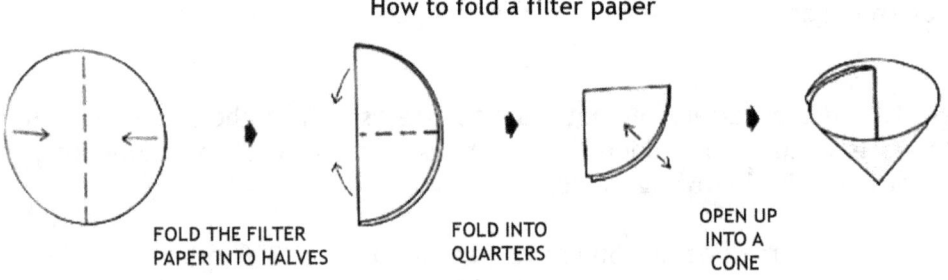

FOLD THE FILTER PAPER INTO HALVES → FOLD INTO QUARTERS → OPEN UP INTO A CONE

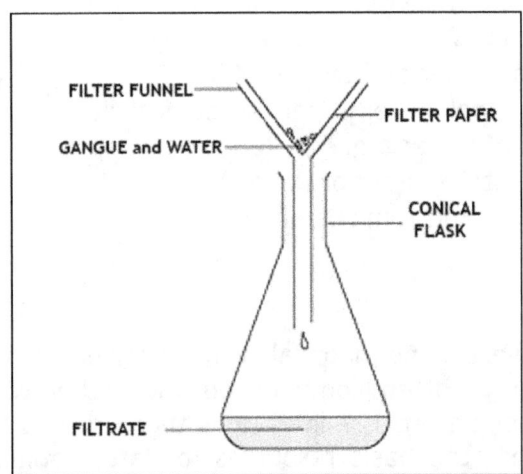

1. Ensure that after the filter paper cone is place in the filter funnel that it is then wet (why?). Also make sure that only a small amount of water-gangue mixture is poured into the funnel so that it does not overflow the top of the filter paper. Remember to change the filter paper if more filtration steps are needed.

2. The first step in the process will be to weigh the initial amount of dry mixture as later on the weights of the dry resources will be weighed so that a yield percent can be found. This is done by:

$$\frac{\text{WEIGHT of DRY RESOURCE}}{\text{WEIGHT of DRY MIXTURE}} \times \frac{100}{1} \quad \%$$

3. Before writing down the Procedure, check that all possible errors are reduced and that tidiness and cleanliness of the equipment and room are considered. Do NOT put waste sand or ore down the drain! Collect all waste in a bucket for later disposal.

EXPERIMENT 20.1 continued

RESULTS:

Weight of initial dry mixture =   g
Weight of dry iron =   g
Weight of dry galena =   g
Weight of dry salt =   g

Sketch the filtration apparatus and make notes about each phase of the separation process giving reasons why the method was used.

QUESTIONS:

1. Why is the mixture used in this activity as ore an unrealistic combination?

2. Why is minimal water to be used in any part of the processes, especially panning?

3. During panning, why should the water and solid mixture be stirred?

4. Why was the filter paper wet before that part of the process?

5. What errors could occur at each phase of the separation process? Suggest any improvements to limit errors.

CONCLUSION:

1. Give a detailed analysis as a yield % for each of the desired resources.

2. Comment how differences in physical properties have been used in this activity of separation of useful materials.

RESEARCH (Optional)

Use the Internet to find out how salt is harvested and purified. Why has salt, mined as halite, been an important resource?

EXPERIMENT 20.2

Time: one lesson

FROTH FLOTATION

AIM: To use simple laboratory materials to demonstrate the principle of froth flotation separation.

MATERIALS: Large test-tube with rubber stoppers, iron filings, sand, detergent, kerosene.

BACKGROUND:

Many ores often are found together and when mined must be separated from each other and from the gangue materials which come with them. The ores of zinc (sphalerite) and lead (galena) often are found mixed together when mined. Froth flotation is a separation technique which can separate both ores using chemicals and an agitator to manufacture bubbles. After crushing to a very fine powder, the ores are placed into vats and a chemical called a surfactant is added to make one of the ores hydrophobic i.e. is repelled by water. Sodium ethyl xanthate is a chemical which will make galena hydrophobic and so when the mixture is run into vats with water and aerated, the galena will rise up with the bubbles and can be skimmed off from the top, dried and collected. Other chemicals can be added to selectively obtain the sphalerite instead.

PROCEDURE:

1. Place a spoonful of the mixture of sand and iron filings into a large test-tube. The sand represents the gangue and the iron filings the ore.

2. Add about 5 ml of water, stopper the test-tube and shake vigorously. Observe what happens when the liquid settles.

3. Add 5 ml of kerosene to the mixture, stopper and shake vigorously. Observe what happens.

4. Now add a couple of drops of detergent, stopper the test-tube and shake vigorously. This represents the complete system. Observe what happens.

RESULTS:

Record the observations in a table for each of the three actions above. Draw a sketch of the final test-tube and its contents labelling the various components seen inside the tube, especially the location of the iron filings and the sand.

QUESTIONS:

1. What is the purpose of the kerosene?

2. What is the purpose of the detergent?

3. How could the iron filings be collected from the test-tube?

4. Why is such a technique important to the mining of metals?

EXPERIMENT 20.2 continued

CONCLUSION:

Summarize the method and observations made during this experiment.

Also see an animation explaining how the technique works at:

https://www.youtube.com/watch?v=pFbiatPD4ZM

RESEARCH (Optional)

Use the Internet to find out how froth flotation is used in other industries.

EXPERIMENT 20.3 Time: one lesson

CHEMICAL EXTRACTION of COPPER from COPPER ORE

AIM: To obtain pure copper from a copper ore.

MATERIALS: Crushed copper ore (azurite/malachite or copper carbonate and sand mixture), dilute sulfuric acid (H_2SO_4), small 250 ml beaker, glass stirring rods, fine sandpaper, filter paper/filter funnel & conical flask, clean iron nails, electronic balance, paper towel, tongs.

BACKGROUND:

Copper can be found in several ores and even as Native Copper. One copper ore is malachite/azurite, a mixture of two different crystal forms of the one complex chemical, copper carbonate – copper hydroxide (usually called basic copper carbonate). Malachite is the green form and Azurite is blue. Both minerals are usually found together and have been deposited from hot water solutions in veins and porous rocks. Chemical extraction of the metal as a solution is often the first step in many processes. As a mixed carbonate, malachite/azurite can be dissolved in acid to give a solution of the copper and the acid. If another metal which is more reactive than copper (e.g. iron) is placed in this copper solution, the copper will be replaced by the active metal and will be seen as a coating on that metal. This coating can be removed as pure copper.

PROCEDURE:

1. Carefully weigh out a small amount of crushed ore (or copper carbonate and sand if the ore is unavailable) – enough to cover the bottom of the small beaker to about 1 centimetre.

2. Add about 50 mL of dilute sulfuric acid (CARE) to the beaker and gently shake the it to ensure that the reaction is complete.

3. Observe and describe any reaction. Wait until all reaction ceases, then add about 50mL of water to dilute any acid which remains.

4. Stir the mixture and filter it so that only transparent copper solution comes through the filter paper (if contaminated and not transparent, re-filter). Filter the mixture as in Expt. 20.1. Sketch the apparatus in the Results.

5. Weigh out a number of iron nails which have been previously sandpapered to remove any surface corrosion. Say about 10 nails and therefore derive a value for the weight of one nail.

6. Add enough iron nails a few at a time to the flask until the colour of the solution disappears. Note any other change in the flask.

7. Carefully pour off (i.e. decant) the solution in the flask and tip the nails onto clean paper towel and allow them to dry.

8. When dry, carefully weigh the nails on the balance. Note any increase in weight for that specific number of nails and hence calculate the amount of pure copper plated on them.

9. Calculate the percentage yield of copper using this value and the initial weight of the ore.

EXPERIMENT 20.3 continued

RESULTS:

Weight of dry crushed ore (or copper carbonate + sand) =   g.

Describe the reaction between the ore and the sulfuric acid. Write a word equation for this reaction assuming the chemical formula for the ore is simply copper carbonate).

Describe the nature of the filtrate which comes through the filter paper by sketching the filter apparatus and indicating the colour of the filtrate.

Weight of iron nails (number =   ) =   g, therefore weight for one nail =   g.

Describe the reaction involving the nails and write a word equation for this reaction.

Weight of copper-coated iron nails   =   g

Weight of copper produced = weight of copper-plated nails − weight of iron nails added to the flask   =   g.

Percentage yield is calculated from: $\dfrac{\text{Weight of Copper produced}}{\text{Weight of ore sample}} \times \dfrac{100}{1}$

QUESTIONS:

1. What is property of the azurite/malachite ore which enables it to be separated from the gangue mixed with it?

2. What is the purpose of the iron nails?

3. What is the chemical property of iron which enables this to happen? What other metals could be used instead of iron? (Hint: see the reactivity series of metals)

4. Why is such a technique important to refining of copper?

CONCLUSION:

Summarize the method and observations made during this experiment.
What is the percentage yield produced by this experiment?
What are some of the probable errors which have occurred? How could they be reduced?

RESEARCH (Optional)

Use the Internet to find out how copper ore is processed and the copper metal refined.

Chapter 21: Monitoring and Management

EXPERIMENT 21.1   Time: one lesson (set up) - extended

MONITORING of LOCAL CLIMATE

AIM: To observe, measure and monitor over time the main variables of climate such as air temperature, air pressure and humidity.

MATERIALS: thermometer, aneroid barometer, wet-dry bulb hygrometer

BACKGROUND:

Most laboratories and environmental scientific institutions keep daily records of the local climate. This can be done electronically and remotely but often is more simply done by manual observations of instruments housed outside in a special box called a Stevenson Screen. In more simple circumstances, the instruments can be set up outside in a shady area away from human movement, air-conditioning and any other man-made influence. This activity is an extension of Practical 11.3 in Practical Manual 1.

PROCEDURE:

1. Set up a weather station of the main instruments listed above, in a secure place under cover and away from direct sunlight. Some places have a proper Stevenson's Screen or electronic weather station equipped for regular weather measurement. If a wet-dry bulb hygrometer is not available, make one using two thermometers attached to light board so that one has a bulb covered in surgical gauze which is dipped in a small container of water.

2. Over a month, measure the parameters of the weather either in class time, or at a set period of day. This may be done on an individual or class group roster. Plot these parameters onto a table/chart.

3. After a month, draw graphs of temperature, relative humidity and pressure then extrapolate (i.e. extend the end of the graph plot) these graphs to predict what the weather will be like in the next few days.

4. Compare this monthly value to the same month in past years by opening the website at:

    http://www.bom.gov.au/climate/data/

(This is for Australian data only. Click on nearest location and then go to the bottom left and select the year. Note the average temperature for the month and go to the next year)

Other places can be researched at: http://www.weatherbase.com/ or

http://www.worldclimate.com/

EXPERIMENT 21.1 continued

DATA and OBSERVATIONS:

Use a table to record each parameter daily e.g.

If measured in proper units using scientific versions of instruments, complete graphs for temperature, relative humidity and air pressure. The following chart will allow a calculation of the relative humidity from the wet and dry bulb hygrometer:

Wet-and-Dry Bulb Thermometer Relative Humidity

Dry Bulb Temp.	Dry Bulb Temperature minus Wet Bulb Temperature (zero difference =100% relative humidity)													
	1°C	2°C	3°C	4°C	5°C	6°C	7°C	8°C	9°C	10°C	12°C	14°C	16°C	18°C
10°C	88%	77%	66%	55%	44%	34%	24%	15%	6%					
11°C	89%	78%	67%	56%	46%	36%	27%	18%	9%					
12°C	89%	78%	68%	58%	48%	39%	29%	21%	12%					
13°C	89%	79%	69%	59%	50%	41%	32%	22%	15%	7%				
14°C	90%	79%	70%	60%	51%	42%	34%	26%	18%	10%				
15°C	90%	80%	71%	61%	53%	44%	36%	27%	20%	13%				
16°C	90%	81%	71%	63%	54%	46%	38%	30%	23%	15%				
17°C	90%	81%	72%	64%	55%	47%	40%	32%	25%	18%				
18°C	91%	82%	73%	65%	57%	49%	41%	34%	27%	20%	6%			
19°C	91%	82%	74%	65%	58%	50%	43%	36%	29%	22%	10%			
20°C	91%	83%	74%	66%	59%	51%	44%	37%	31%	24%	11%			
21°C	91%	83%	75%	67%	60%	53%	46%	39%	32%	26%	15%			
22°C	92%	83%	76%	68%	61%	54%	47%	40%	34%	28%	16%	5%		
23°C	92%	84%	76%	69%	62%	55%	48%	42%	36%	30%	19%	7%		
24°C	92%	84%	77%	69%	62%	56%	49%	43%	37%	31%	20%	9%		
25°C	92%	84%	77%	70%	63%	57%	50%	44%	39%	33%	22%	13%		
26°C	92%	85%	78%	71%	64%	58%	51%	46%	40%	34%	23%	14%	4%	
27°C	92%	85%	78%	71%	65%	58%	52%	47%	41%	36%	26%	16%	7%	
28°C	93%	85%	78%	72%	65%	59%	53%	48%	42%	37%	27%	17%	8%	
29°C	93%	86%	79%	72%	66%	60%	54%	49%	43%	38%	29%	20%	11%	
30°C	93%	86%	79%	73%	67%	61%	55%	50%	44%	39%	30%	20%	12%	4%
32°C	93%	86%	80%	74%	68%	62%	56%	51%	46%	41%	32%	23%	15%	8%
34°C	93%	87%	81%	75%	69%	63%	58%	53%	48%	43%	34%	26%	18%	11%
36°C	93%	87%	81%	75%	70%	64%	59%	54%	50%	45%	26%	28%	21%	14%
38°C	94%	88%	82%	76%	71%	65%	60%	56%	51%	47%	38%	31%	23%	17%

For example: The relative humidity with dry = 24°C and wet = 20°C is 69% at 24°C

RESULTS:

1. Set up a chart for the month (or other time periods) showing the daily air temperature (wet and dry if needed), relative humidity and air pressure e.g.

DAY (Month)	AIR TEMPERATURE		REL. HUMIDITY %	AIR PRESSURE hectopascals (hPa)
	Dry Bulb	Wet Bulb		

2. Draw graphs for each of the parameters over the month and note any changes.

3. Draw a graph for the local average temperature over time (e.g. 1945 to present every 5 years)

QUESTIONS:

1. Has the weather changed over the month? How (give details).

2. Was the weather in the next week as predicted? Comment.

3. Has the average for that month temperature changed locally over time? (give details).

EXPERIMENT 21.1 continued

4. List all of the problems or errors which could occur with such short-term weather measurement.

CONCLUSIONS:

Comment on the usefulness of monitoring local climate and any changes observed over the (a) short term and (b) long term.

RESEARCH (Optional):

1. How is weather data gathered on a wider scale?

2. How are weather predictions made?

EXPERIMENT 21.2   Time: one lesson and extended

MONITORING of CARBON DIOXIDE LEVELS

AIM: To observe, collect data and monitor over time the levels of atmospheric carbon dioxide.

MATERIALS: Test-tubes, test-tube racks, drinking straws, clear saturated limewater (calcium hydroxide solution), rubber stoppers for test tubes, computers or tablets, internet connection.

BACKGROUND:

Carbon dioxide gas is odourless, colourless and the current atmospheric concentration is about 0.04% or 410 parts per million (ppm) by volume. It is produced by respiration of living things and by carbonate reactions in the lithosphere. It is also absorbed by green plants for photosynthesis and by the oceans. Carbon dioxide is very soluble and dissolves in the ocean to form carbonic acid (H_2CO_3) and bicarbonate (HCO_3^-) and carbonate (CO_3^{2-}) ions. There is about fifty times as much carbon dissolved in the oceans as exists in the atmosphere and the oceans act as an enormous carbon sink, and it has been estimated that the oceans have taken up about a third of CO_2 emitted by human activity so far. The gas is also dangerous to Humans in higher concentrations and in concentrations up to 1% it will make some people feel drowsy as in a crowded, closed room. At 7% to 10% it may cause suffocation, even in the presence of sufficient oxygen, giving dizziness, headache, visual and hearing dysfunction, and unconsciousness within a few minutes to an hour. At 1 atmosphere, the gas freezes directly to a solid below –78.5 °C (–109.3 °F; 194.7 K) to dry ice.

PROCEDURE:

Part A: Carbon Dioxide in the Room

1. Pour clear limewater into three test-tubes to about 3 cm depth.

2. Stopper one test-tube and put it and a second test-tube into the racks and stand it aside in a secure place. Observe any changes at the end of the lesson and over several days.

3. Insert the drinking straw into the limewater of the third test-tube and blow into it _gently_. Compare the result with that of steps 1 and 2.

RESULTS:

1. Record any changes for steps 2 over several days in a table.

2. Compare the results of step 3 with step 2 during the lesson.

QUESTIONS:

1. Why was one of the test-tubes stoppered?

2. What did the results of step 2 (unstoppered) and step 3 show?

3. Was there any change to the test-tubes in the test-tube rack over several days? Explain.

EXPERIMENT 21.2 continued

CONCLUSIONS:

Comment on the amount of carbon dioxide in the breath due to respiration compared to that in air. Also give an hypothesis for any change observed over several days. Comment on the usefulness of this method as a way of detecting carbon dioxide gas in other locations.

PROCEDURE:

Part B: Carbon Dioxide Levels in the Atmosphere Worldwide

1. Open the following website which gives daily CO_2 readings from the observatory at Mauna Loa, Hawaii:

    https://www.co2.earth/daily-co2

    Bookmark this site so that regular readings could be taken in the future.

2. Open the following website which gives carbon emissions data from the past:

    https://data.giss.nasa.gov/modelforce/ghgases/Fig1A.ext.txt

    Use the first set of data (four columns) from ice core mixes.

3. Graph the CO_2 levels (ppm) against years from 1940 to 2010 in increments of 10 years.

4. Use the graph to extrapolate and find the probable level of carbon dioxide in 2030.

RESULTS:

Record daily carbon dioxide level over one month.

Graph of carbon dioxide gas levels from 1940 with extrapolation.

QUESTIONS:

1. What is the current level of carbon dioxide gas at Mauna Loa in parts per million? How does this agree with the extrapolation? Explain any differences (+/- 10%).

2. Why are the values taken from measurements at Mauna Loa in Hawaii?

3. Was there any change in the recorded CO_2 levels over the month? Explain, giving details of any possible errors in sampling.

CONCLUSION:

Make a general comment on the change in CO_2 levels over the last sixty years and give some explanation for any changes.

RESEARCH: (Optional) How and where are these measurements made and what errors could be involved in measuring carbon dioxide levels?

EXPERIMENT 21.3   Time: one lesson and extended

MONITORING of AIR QUALITY

AIM: To observe, collect data and monitor over time the levels of air quality in the local area and some selected international centres.

MATERIALS: Computers or tablets, internet connection.

BACKGROUND:

With the advent of industrialisation, large volumes of industrial gases and particles have been put into the atmosphere. For many reasons, especially for the health of local people and livestock, many countries have become concerned about the level of global air pollution. An air quality index (AQI) is a number used by governments to the public how polluted the air currently is or how polluted it is forecast to become. As the AQI increases, an increasingly large percentage of the population is likely to experience increasingly severe adverse health effects. Different countries have their own air quality indices, corresponding to different national air quality standards. The concentration measured include pollutants such as sulfur dioxide (SO_2), nitrogen dioxide (NO_2), carbon monoxide (CO), ozone (O_3) and various sizes of particulate matter (e.g. $PM_{2.5}$ particles less than 2.5. micrometres and PM_{10} less than 10 micrometres).

PROCEDURE:

Part A: Local and national air quality

1. Open the following website for Australian air quality values:

    http://aqicn.org/here/   (Australia)

2. Scroll and left-mouse drag the map and notice some of the measuring sites. Click on some of these sites to see the current values of air quality. Note that not all of the localities are shown on the map. Use the Search Box (top right) to type in a locality which may have to be the nearest major centre.

3. Go to the nearest local centre using the Search Box and use a table to record the values for the time and date and current values for the AQI, temperature, air pressure and the AQI levels for PM 2.5, PM 10, ozone (O_3), nitrogen dioxide (NO_2) sulfur dioxide (SO_2) and carbon monoxide (CO).

4. Set up the table as a chart so that these values can be recorded over a time period e.g. over the day at several times, a week or a month (choose whichever is practical):

Date/time	AQI	Temp.	Pressure	PM 2.5	PM 10	O_3	NO_2	SO_2	CO

etc.

EXPERIMENT 21.3 continued

RESULTS:

Chart of values for AQI and individual pollutants (as above).
Any general comments about trends or changes over the time.

QUESTIONS:

1. What was the value for the AQI first measured? How was it rated?

2. How has it changed over time? Go to the locality site and look at the values given for the last 48 hours and comment on any changes.

3. Have there been any climatic change during the last 48 hours to cause these changes? (e.g. rain, dust storm, fires etc.)

CONCLUSIONS:

Make any comment generally about the local AQI giving reasons why it should have such a rating. Also comment on any local conditions of climate or other factors which might cause changes to the local AQI.

RESEARCH: optional

1. Other than bookmarking this website, where can the local daily AQI value be found?

2. Is the local air quality subject to changes over time?

Part B: International values of AQI

1. Open the website and select several different localities around the world, choosing some of interest and others which are known to have variable or high values of AQI e.g. Sydney, Los Angeles, London, Noumea, Wellington, Oslo, Tokyo, Mumbai, Beijing, Berlin, Cairo, Moscow etc. Choose at least 10 sites to show some variation of AQI values.

    This site may be used as an alternative to get some ideas about selection:

    https://waqi.info/

2. Draw up a table as in Part A but changing the date/time column to a locations column into which the ten location names are inserted.

RESULTS:

Chart of values for AQI and individual pollutants for ten locations.
Any general comments about values and locations including possible reasons for such values.

EXPERIMENT 20.3 continued

QUESTIONS:

1. What locations had (a) high values and (b) low values of AQI?

2. What were the main individual AQI pollutants in places with high values? Give probable reasons for these high values.

3. What factors would vary such high values from time to time?

CONCLUSIONS:

Make any comment generally about the international AQI values and reasons for major differences observed across world locations such as climate or other factors.

RESEARCH: (Optional)

1. How and where are AQI values calculated? What are their usefulness and limitations?
2. Name some places which have extremely high and dangerous values of AQI?

EXPERIMENT 21.4

Time: one Lesson

RADON GAS LEVELS

AIM: To use an interactive map which allows the acquisition of radon gas emissions nationally.

MATERIALS: Computers or tablets and Internet access to the website for Australian radon concentrations at:

http://arpansa.maps.arcgis.com/apps/Embed/index.html?webmap=c7501ea15f45467da37059b21ec8e66e&extent=105.5019,-42.4303,167.2451,-11.9624&home=true&zoom=true&scale=true&search=true&searchextent=true&legend=true&basemap_gallery=true&theme=light

(Website is from the Australian Radiation Protection and Nuclear Safety Agency)

BACKGROUND:

Radon is a radioactive, colourless, odourless and tasteless unreactive monatomic gas belonging to the noble gas family of elements. It occurs naturally as an intermediate step in the normal radioactive decay chains of the elements, thorium and uranium. Its most stable isotope, ^{222}Rn, has a half-life of only 3.8 days but as it comes from elements with very large half-lives, it is constantly being renewed. Radon gas is a dense gas being eight times heavier than air so it sinks into low-lying places, especially the basement of buildings in urban areas. It is thought to be the second most common cause of lung cancer after cigarette smoke in urban areas. Radon concentration in the atmosphere is usually measured in becquerels per cubic metre (Bq/m^3), the SI derived unit. The average concentration of radon gas ranges from less than 10 Bq/m^3 to over 100 Bq/m^3 in some European countries and the United States Environmental protection Agency has adopted a radon concentration of 200–400 Bq/m^3 for indoor air as an upper reference level for health with up to 150 Bq/m^3 requiring no action.

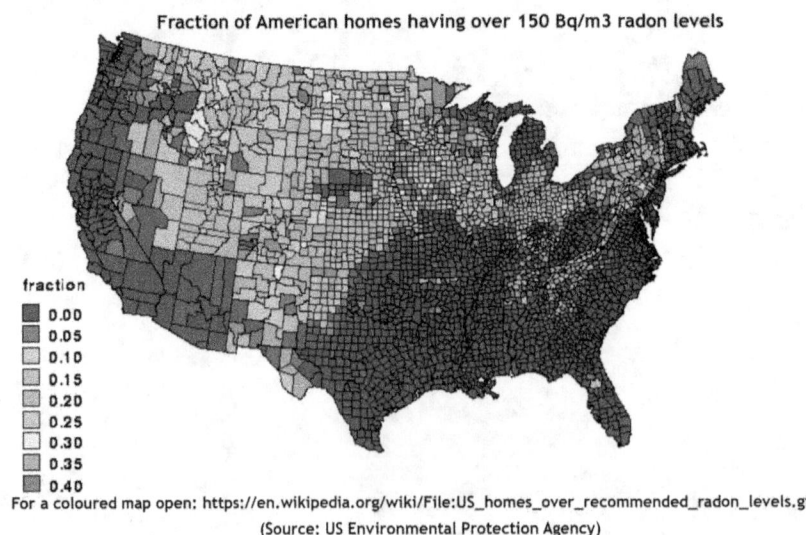

Fraction of American homes having over 150 Bq/m3 radon levels

For a coloured map open: https://en.wikipedia.org/wiki/File:US_homes_over_recommended_radon_levels.gif

(Source: US Environmental Protection Agency)

EXPERIMENT 21.4 continued

PROCEDURE:

1. Go to the website and examine the general patterns of concentrations on the map.

2. Find the local area on the map or type in its name in the box at the top right-hand corner. If no value appears, use the mouse to Left Click the nearby region. Note this value in the Results.

3. Go to each major urban area in Australia (e.g. Sydney, Melbourne, Brisbane, Perth, Adelaide, Gold Coast Queensland, Newcastle NSW and Canberra ACT) and click on their locations to obtain a reading for Australian urban centres. Record these in a table in the Results and find an average for the Australian Urban Radon Concentration.

4. Re-examine the map and locate regions which have a higher concentration of radon emission. Record some of these values and compare these areas with the Australian Surface Geology map at:

    https://d28rz98at9flks.cloudfront.net/73360/Geology_A3.pdf

RESULTS:

1. Record the local concentration of radon emissions.

2. Construct a table for the radon concentrations in the local and main urban centres:

Location	Radon Concentration (Bq/m^3)	Comments
Sydney etc.		

3. Average for Australian urban centres

4. Greatest values for some Australian locations (record some of the values in the yellow and light green regions of the map).

QUESTIONS:

1. How does the local value of radon emissions compare with that of the average for Australian urban centres?

2. Comparing the maps for the United States and for Australia, what can be said about the average radon emissions for each country?

3. Where are the major places for greatest radon emissions in Australia?

EXPERIMENT 21.4 continued

QUESTIONS: (continued)

4. Examine the map of Australian surface geology by looking carefully at the legend at bottom left of the map and then zooming in to the areas of higher radon concentration (using the + feature at the top menu bar). Is there a geological correlation or match between rock type and/or geological structure and higher concentrations of radon emission?

5. Locate Olympic Dam in South Australia by typing its name into the search box. It is the largest underground mine of uranium (a source of radon gas) in Australia and fourth largest in the world. Its radon emission levels are about 12-13 Bq/m^3 and seems low compared to that of the nation's capital, Canberra which is about 15-16 Bq/m^3. Why?

6. Australia has a relatively low average radon emission level compared to other countries with the world average of 26 major counties being 64.3 Bq/m (World Health Organisation data at: https://www.who.int/ionizing_radiation/env/radon/IRP_Survey_on_Radon.pdf). Why is Australia's emission low compared to the average?

CONCLUSION:

Write a general conclusion about the average Australian emission levels of radon gas and where higher concentrations are found and why. In your answer, also give a judgement about the health hazards of radon emissions in Australia compared to other countries.

(see also: https://www.thenakedscientists.com/forum/index.php?topic=71426.0)

RESEARCH: (Optional)

1. Research the life and work of Henri Becquerel.

2. What workplace health and safety issues are there in mining and prospecting for uranium?

EXPERIMENT 21.5   Time: one or two lessons with possible extension

WATER QUALITY MONITORING

AIM: To monitor part of the hydrosphere by testing collected water samples.

MATERIALS: Collected water samples (at least 5 different sources), commercial water sampling kit or separate laboratory equipment and reagents, large test-tubes, Universal Indicator solution and chart, indigo carmine solution, 250 ml conical flask, soap, distilled water, test-tube rack, silver nitrate solution, barium nitrate solution iron III sulfate solution, conc. sulfuric acid.

Water samples can be collected in plastic bottles which have been thoroughly washed, rinsed and dried before use. About 500 to 1000 ml is needed for each sample.

BACKGROUND:

Monitoring the hydrosphere is a major activity of the environmental Earth scientist. Water in natural settings is monitored to ascertain the health of the ecology. Water in the urban setting is analysed for its quality and suitability for drinking and in the rural setting for its suitability for livestock and crops. Water is also monitored in an industrial setting for its suitability for the industrial process and before it is discharged into the environment. In mining, water is constantly monitored for toxic wastes before being returned to the environment. Water monitoring can be done directly in the field using portable monitoring kits and digital meters or water can be sampled in the field and tested in the laboratory under more suitable conditions. Water monitoring can also be done from place to place to ascertain the general health of the environment during a single survey, or a specific body of water can be monitored over time as part of a regular and continual monitoring process.

There are some test strip kits which can be used in testing saltwater and other pools, but some of these are expensive but they do give results. Some of the chemicals below may not be sensitive enough to detect the pollutant but they show the basic principle of the test.

PROCEDURES:

Carry out the following water tests noting the object of the test, the method used and the result. If there is no positive result this may be demonstrated for you:

PART A: pH
This is a measure of the acidity or alkalinity of the water. High values or acidity or alkalinity could indicate the presence of pollution or an imbalance of the ecology. Organisms have different pH tolerances but most seem to be around that of neutral conditions at pH = 7 (+/- 1.0). Some extremophiles can live in conditions well outside of normal values.

1. A small sample (about 2 cm) of the water is placed into a test tube and a few drops of Universal Indicator are added.

2. Shake the test tube and compare the pH to a standard pH colour chart supplied.
    a. (or see https://www.rapidonline.com/edu-resources/docs/ph-scale.pdf)

3. Record this value in a table for the whole water test under pH.

EXPERIMENT 21.5 continued

PART B: Water hardness

Some natural chemicals in the water make the water <u>hard</u> to lather with soap – important for washing and several industrial processes as hard water often leaves a scale of these minerals on the insides of pipes and equipment. Temporary hardness is caused by the presence of dissolved bicarbonate of calcium e.g. Ca $(HCO_3)_2$ and magnesium e.g. Mg $(HCO_3)_2$ and can be removed by boiling. Permanent hardness is caused by the presence of sulfate (SO_4^{2-}) and chloride (Cl^-) ions of calcium and/or magnesium and cannot be removed by boiling. A water conditioner such as Washing Soda (sodium carbonate - Na_2CO_3) must be added to remove these ions.

1. Add about 2 cm of water sample to a test-tube and to another, separate test-tube add the same amount of distilled (de-ionized) water to act as a control.

2. Add a small pellet (about pea size) of pure soap to each test-tube, cover with a thumb and vigorously shake both test-tubes to see if a lather appears (= no hardness).

PART C: Test for Chlorides (Demonstration - silver nitrate stains!)

1. Add about 2 cm of sample water to a test-tube and add two drops of silver nitrate solution.

2. Observe closely against a dark background to see if there is any cloudiness due to the reaction of the chloride ions to form white, insoluble silver chloride.

Alternatively, a Chlorine Test Strip used in testing saltwater pools can be used for determining the free chloride content as well as the pH (see: http://hugohd.com/editor/)

PART D: Test for Sulfates

1. Add about 2 cm of sample water to a test-tube and add a few drops of barium nitrate solution.

2. Observe closely against a dark background to see if there is any cloudiness due to the reaction of the sulfate ions to form white, insoluble barium sulfate.

PART E: Oxygen content:

1. Add 4-5 ml of fresh Indigo Carmine solution to about 100 ml of the water sample in a conical flask and mix with a gentle shake.

2. Compare this to the colours below to find the oxygen level in parts per million (ppm):
   Yellow 0.000
   Orange 0.005
   Orange-Pink 0.010
   Pink 0.015
   Pink-Red 0.025
   Red-Purple 0.050
   Purple 0.100

   (also see: https://www.youtube.com/watch?v=_e8ENtdBmlc)

EXPERIMENT 21.5 continued

Part F: Nitrates - Demonstration of the Brown Ring Test

1. Add a small amount of sample water to a test tube and add some iron III sulfate solution (saturated) so that the test tube is only half filled.

2. Slowly add drops of conc. sulfuric acid (CARE) to the test tube which is slightly tilted so that the denser acid will slide down the inside and go below the solution of the nitrate. A brown ring will form at the junction of the two layers, indicating the presence of the nitrate ion. This test is sensitive up to a concentration of 1 in 25,000 parts.

This test may not be sensitive enough for natural water. There are also pool test strips for nitrogen ions and ammonium ions in solution but these are expensive.

RESULTS:

Record the results of each test in a table e.g.

SAMPLE	SOURCE	pH	HARDNESS	Cl⁻	SO_4^{2-}	O_2	NO_3^{2-}

Give any colours and values as able or a simple YES/NO for the presence of the substance.
Also make any other observations as appropriate

QUESTIONS:

1. Why was the sample bottle originally washed, rinsed and dried before collecting water?

2. What is the usual pH level preferred by most organisms?

3. Where do the calcium, magnesium, hydrogen carbonate, sulfate and chloride ions in water come from?

4. What are some of the main errors in the methods used here for water sampling?

CONCLUSION:

Make a list showing the name of the test, what property/substance it tests for and the result if positive.

Discuss the need and usefulness of water testing for (a) domestic water supplies, (b) the local ecology, (c) industry and (d) mining (some research may be needed).

RESEARCH: (Optional)

Research how water quality is monitored in the oceans See NOAA sites)

EXPERIMENT 21.6   Time: one lesson

TOTAL DISSOLVED SOLIDS - DEMONSTRATION

AIM: To show how Total Dissolved Solids are measured during water monitoring projects.

MATERIALS: Seawater or made-up saltwater, large measuring cylinder, large spherical flask, electric balance, mat, tripod, gauze, Bunsen burner, multimeter, small rectangular but deep dish (e.g. a small baking dish), wire loop, hand spectroscope (optional), copper or aluminium foil.

BACKGROUND:

Total Dissolved Solids (TDS) is the sum of the cations (positively charged) and anions (negatively charged) in water and provides a qualitative measure of the amount of dissolved substances in the water. TDS monitoring is important in the processing of drinking water and in many industrial processes where dissolved materials could cause harm to the equipment if they solidify. TDS in drinking-water can originate from natural sources, urban run-off, industrial wastewater, and chemicals used in the water treatment process, and the nature of the piping or hardware used to convey the water. Whilst salinity is treated as a separate parameter of water, saltwater in this experiment is used as a matter of convenience.

There are two ways of measuring TDS; gravimetric analysis involving the physical removal of the water content and the weighing of the residue; and electrical conductivity which is an estimate (+/- 10%) using the fact that conductivity of water is directly related to the concentration of dissolved ionized solids in the water.

PROCEDURE:

PART A: Gravimetric Analysis

1. Weigh a large empty flask on an accurate electronic balance and then add at least 500 ml of water sample. This must be an accurately-measured amount.

2. Place it on a strong tripod with gauze on a heat-proof mat and heat the flask strongly with a Bunsen burner to boil off all of the water (This may take some time so the second part of the experiment can be done).

3. When the contents of the flask have boiled dry, allow it to cool then reweigh on the balance.

4. Calculate the amount of Total Dissolved Solid by subtracting the original empty weight of the flask from this last value.

5. Some of the solid can be scraped out using a wire loop (e.g. a paper clip unfolded) and put into the blue flame of the Bunsen to observe the coloured flame as a simple form of spectroscopic analysis (or the flame could be viewed through a hand-held spectroscope for an accurate verification).

EXPERIMENT 21.6 continued

RESULTS:

(1) Weight of empty flask   =   g.

(2) Volume of sample water added = ml.

(3) Weight of flask + dry TDS = g.

(4) Weight of dry TDS = (3) - (1) = g.

Time taken for complete removal of water =   minutes.

QUESTIONS:

1. Why was the water sample volume measured accurately?

2. What was the final weight of the TDS?

3. What was the original concentration of the solid? (This is found by calculating how much solid would be in one litre of water)

4. What are some of the main errors or disadvantages in using this method?

PART B: Conductivity estimation of TDS

1. Pour a sample of the test water into the dish so that there is a good depth - say about 4 cm.

2. Turn on the multimeter and set to the it to the Resistance setting (Ω) - at maximum value e.g. 2000 Ω.

3. Touch the electrodes of the meter to the water at each of the long ends of the dish and measure the <u>resistivity</u> (note: this is the reciprocal of conductivity) e.g. say 70 ohms.

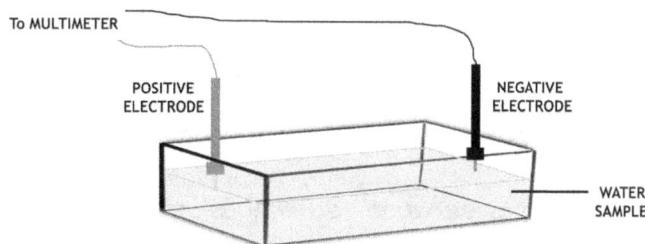

4. Measure the length, width and depth of the glass dish in centimetres e.g. consider a length of 22 cm, a width of 13 cm and a depth of 4 cm.

5. Multiply the width by the depth to obtain the area of the ends of the dish in square centimetres e.g. 13 cm x 4 cm giving an area of each end of 52 square centimetres (cm^2).

EXPERIMENT 21.6 continued

6. Divide the length (e.g. 22 cm) by the product of the resistance and the area of an end (e.g. 70 ohm x 52 cm²) to arrive at the conductivity in units of siemens per meter (i.e. 22 /70 x 52 = 0.006). This gives the conductivity in siemens per meter (i.e. 0.006 S/m).

7. Convert the conductivity to microsiemens per centimetres (µS/cm) by multiplying by 10,000. One microsiemen is one-millionth of a siemens e.g. 10,000 x 0.006 = 60 µS/cm.

8. Use the value for conductivity calculated from the water sample in this experiment to estimate the TDS by the formula (correct saltwater at 20 °C):

    TDS (milligrams/Litre) = Conductivity (µS/cm) X 0.7

    From the example used as a demonstration, the TDS would be equal to:

    60 µS/cm x 0.7 = 42 milligrams/Litre (mg/l)

RESULTS:

Depth of water in dish = cm. Length of dish = cm. width of dish = cm.

Value for resistance = ohms.

Calculated value for Conductivity = µS/cm

Calculated value for TDS (estimate) = mg/l

(if the saltwater solution was made up then the real concentration was mg/l)

QUESTIONS:

1. What was the instrument error of (a) the multimeters, (b) the measured dimensions of the tray and (c) the area of one end?

2. What other errors could occur in this estimation (itself a very rough method)?

CONCLUSION:

1. What was the value for the TDS determined by gravimetric analysis (accurate)?

2. What was the estimated value of the TDS using conductivity?

3. Comment on the differences between both values and any errors involved.

4. Comment of the pros and cons of using both methods

RESEARCH: (Optional)

Use the internet to find out how the TDS (dry) could be analysed to find out its exact composition.

EXPERIMENT 21.7   Time: one lesson with extension

TURBIDITY MANAGEMENT

AIM: To show some basic principles of how turbidity in water can be managed.

MATERIALS: 2l plastic drink bottles, 100 ml measuring cylinders, filter funnels, flask, filter paper, beakers, sand-silt-gravel mix, flocculant (e.g. aluminium sulfate), mini Secchi disk (metal washers, cardboard, string, plasticine)

BACKGROUND:

Turbidity is caused by particles suspended or dissolved in water that scatter light making the water appear cloudy or murky. Particulate matter can include sediment, especially clay and silt, fine organic and inorganic matter, coloured organic matter such as tannin, algae, and other microscopic organisms. High turbidity can significantly reduce the beauty of lakes and streams, increase the cost of water treatment for drinking and food processing and it can harm fish and other aquatic life. Places where water becomes turbid due to introduced sediments such as on industrial and mining sites, settling ponds are often used to reduce the sediment in the water flowing from the site. In some places, flocculants are added to the settling ponds. These are chemicals such as aluminium sulfate which attract suspended sediment to their particles creating bigger particles which can then settle to the bottom faster.

PROCEDURE:

PART A: Sedimentation and settling.

1. Add a mixture of sand, silt and gravel to a 2l plastic bottle using a funnel so that the sediment fills about the bottom 4 or 5 cm of the bottle.

2. Fill the bottle with water, cap it and then shake the bottle vigorously. This will represent the turbid water within a settling pond or tank.

3. Set it down and watch the sedimentation process. See the following video from a previous Chapter of the textbook: https://www.youtube.com/watch?v=yaYO4lc_G3M

4. Place one bottle aside and allow it to settle overnight. Other bottles will be used for other purposes.

5. From another bottle just shaken, carefully pour off (decant) some of the water and suspended sediment into two identical 100 ml measuring cylinders. This represents running top water from one sediment pond into another.

6. Into one, sprinkle about a spoonful of aluminium sulfate powder and note the time.

EXPERIMENT 21.7 continued

7. Construct mini Secchi disks from cardboard cut in a circle just smaller than the diameter of a 100 ml measuring cylinder. Glue string to its centre using plasticine or putty so that it hangs correctly:

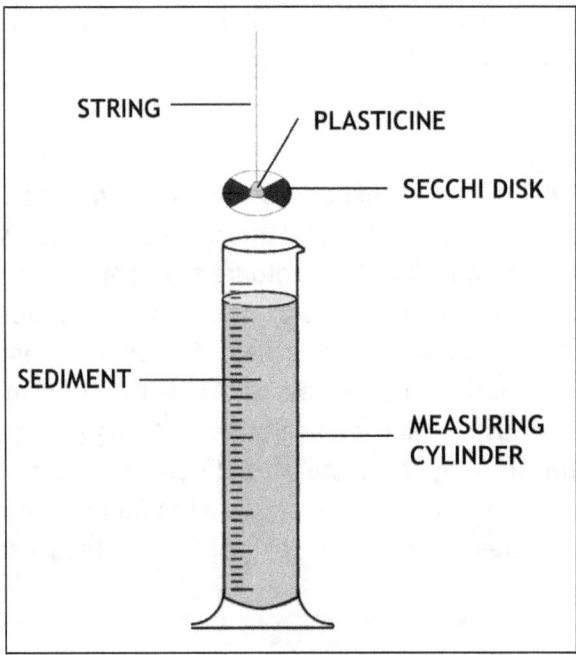

8. Hang the Secchi disk into cylinder and measure the depth which just obscures the pattern of the disk and record this value. Note the time.

9. After a suitable time (say 5 minutes) repeat Step 8. Do this for several time periods to give about five readings in total (Caution: this may take some time).

10. Graph the depth of Secchi disk values against time and extrapolate the graph to find out when the water will be clear enough to see the Secchi disk if placed at the bottom of the cylinder (i.e. the total depth of water in it).

11. Now look at the other measuring cylinder and compare the turbidity of both sediments in the measuring cylinders to see any affects of the aluminium sulfate.

12. The next lesson, compare the turbidity of the bottle which had been set aside to what it was the day before (it can be re-shaken as a reminder)

RESULTS:

Describe how do the sediments settle after the shaken bottle is allowed to stand.

Construct a graph of time v Secchi disk depth and extrapolate the line/curve.

Describe any effect of adding aluminium sulfate.

EXPERIMENT 21.7 continued

QUESTIONS:

1. How did the sediment settle immediately after the shaken bottle was put down?

2. How long did it take for the coarser sediments to settle compared to the fine, suspended sediments?

3. What factors will determine the rate of settling of sediments in water?

4. Did the Secchi dish prove effective in measuring turbidity?

5. What was the estimated time (by graph extrapolation) for the suspended solid to clear?

6. Was there this clarity in the bottle set aside over night? Comment.

CONCLUSION:

Using the bottle as a model, comment on the use of settling tanks to reduce turbidity in waterways and industrial sites.

Comment on the effectiveness of chemicals such as aluminium sulfate (Alum) in reducing turbidity.

RESEARCH: (Optional)

1. Use the internet to find out how sedimentation ponds are constructed.

2. What other methods can be used to remove sediment from water on a large scale?

Chapter 22: Renewable Resources

EXPERIMENT 22.1 Time: one or two lessons

SUSTAINABILITY of TUNA as an ECOSYSTEM RESOURCE

<u>AIM:</u> To examine the potential sustainability of tuna as a food resource

<u>MATERIALS:</u> Computers or tablets with Internet connection, graphing facilities - electronic or paper.

<u>BACKGROUND:</u>

The Atlantic Bluefin Tuna ((*Thunnus thynnus*) is one of the world's most valuable food resources. There are two distinct populations of Atlantic Bluefins: a western population which spawns in the warm Caribbean and migrates out into the colder central Atlantic Ocean; and an eastern population which spawns in the warm Mediterranean Sea and also migrates out into the Atlantic but keeps closer inshore to Europe and Africa. The eastern fish is much larger than the western and also has a much larger population. Since the 1970's new technology in locating these fish and new efficient methods of fishing have resulted in a rapid depletion in catches of both populations.

<u>PROCEDURE:</u>

<u>PART A: Earlier population numbers</u>

Examine the following data which shows an estimate for the western population of Atlantic Bluefin Tuna since the 1970's. Construct a graph using this data but make provision for the time line to go to the present time.

YEAR	BIOMASS TREND (tonnes)
1975	30,000
1980	12,500
1985	10,000
1990	9000

(Data source: International Commission for the Conservation of Atlantic Tunas)

<u>RESULTS:</u>

Draw the graph of population of the western Atlantic Bluefin Tuna.

<u>QUESTIONS:</u>

1. What does this graph show about the population of this fish population?

2. Considering that this fish species has been caught commercially in the Atlantic for over 100 years, what would have given this result? (some Internet research in tuna fishing since 1970 would help)

EXPERIMENT 22.1 continued

PART B: Modern times

3. Add to this graph by completing its pathway for more recent estimations of the population:

YEAR	BIOMASS TREND (tonnes)
1995	10,000
2000	11,000
2005	9000
2010	11,000
2015	19,000

(Data source: International Commission for the Conservation of Atlantic Tunas)

Note: The graph for the eastern Bluefin Tuna is almost identical in shape despite their population being larger.

RESULTS:

Add the above data to the graph previously drawn.

QUESTIONS:

1. What does this graph show about the population of this fish population after 1990?

2. What would have given this result? (some Internet research in tuna fishing since 1990 would help).

3. Is the catching of western Bluefin Tuna now considered sustainable?

4. Open the following website and read the article carefully. Compare and contrast the data shown in Figure 2 for both population for the years (a) 1974 – 1995 and (b) 1995 to 2015.

https://www.pewtrusts.org/en/research-and-analysis/issue-briefs/2017/10/the-story-of-atlantic-bluefin

5. What would be a factor which could lead to some inaccuracy in these graphs (Figure 2)? Explain.
Also see:
https://www.pewtrusts.org/en/research-and-analysis/fact-sheets/2013/10/08/the-story-of-atlantic-bluefin-tuna

6. What caused factors the sudden decrease in trend of population from the 1970's until the mid-1990's?

7. What were the two main methods of large-scale fishing used in catching tuna? (some research may be needed here to explain these methods).

8. What steps have been taken to reverse the overfishing of Atlantic tuna? Is there any cause for optimism that this natural food resource could become sustainable in the future? Why?

EXPERIMENT 22.1 continued

PART C: Another example -Southern Bluefin Tuna

The southern bluefin tuna (*Thunnus maccoyii*) is found in open southern hemisphere waters around Australia, New Zealand and South Africa. At up to 2.5 metres long and weighing up to 260 kilograms, it is among the largest and most valuable fish species. Most countries, except Australia, use the long-line method using a trailing fishing line tagged with hooks. Australian tuna fishermen operating out of southern states use the purse seine method which uses a circular, floating net which drags fish in towards the boat.

1. Open the website at:

    https://www.ccsbt.org/sites/default/files/userfiles/file/docs_english/meetings/meeting_reports/ccsbt_25/Attachment06_from_report_of_SC23.pdf

2. Read the document carefully and then scroll down to the graphs on pages 7 and 8.

QUESTIONS:

1. In figures 1 & 2 give an explanation for the sudden change in catch numbers:

    (a) After 1960?
    (b) From 1960 to 1990
    (c) From 1990 to 2016

2. What changes occurred to the methods of fishing after 1999?

3. Considering the catches over the last 20 years do the graphs show any hope for sustainability of Southern Bluefin Tuna in the future? Discuss.

4. Scroll down to Figure 4. What do these images show about changes in the geographical locations of tuna fishing catches?

5. What attempts have been made to reduce over-fishing and manage Southern Bluefin Tuna as an environmental resource?

The following websites may assist in these questions:

https://theconversation.com/australian-endangered-species-southern-bluefin-tuna-11636

https://www.dpi.nsw.gov.au/__data/assets/pdf_file/0004/508018/southern_bluefin_tuna_sis_part_1.pdf

https://www.ccsbt.org/en/content/about-southern-bluefin-tuna

CONCLUSION:

Write a report about the use of tuna as an ecosystem resource. Comment on demand, uses, fishing catch trends from the 1960's with reasons for these trends, methods of fishing and attempts to make the industry sustainable.

| EXPERIMENT 22.2 | Time: one to three lessons |

SUSTAINABILITY of TIMBER as an ECOSYSTEM RESOURCE

<u>AIM:</u> To examine the potential sustainability of timber as an ecosystem resource using selected algorithms on a world map

<u>MATERIALS:</u> Computers or tablets with Internet connection.

<u>BACKGROUND:</u>

Forest ecosystems are one of the main ways by which the atmosphere receives new oxygen and removes carbon dioxide as well as playing a major role in the Earth's water cycle. Forests and trees play crucial roles in providing food security, drinking water, medicine, renewable energy and other timber resources. They provide around 20 percent of income for rural households in developing countries as well as fuel for cooking and heating for one in every three people around the world. Sustainable forest management attempts to keep the balance between a nation's ecological, economic and socio-cultural development. Successful sustainable forest management will protect the forest biodiversity and ecosystems, reduce rural poverty and mitigate some of the effects of climate change.

<u>PROCEDURE:</u>

1. Open the following interactive website (it will take time to load):

https://www.globalforestwatch.org/map?mainMap=eyJzaG93QW5hbHlzaXMiOnRydWV9&map=eyJ6b29tIjo0LCJjZW50ZXIiOnsibGF0IjotMjIuOTU4MzkzMzE4MDg2MzQ4LCJsbmciOjEyOC40MDgyMDMxMjUwMDAwM30sImNhbGJvdW5kIjpmYWxzZSwiYmJveCI6bnVsbH0%3D&menu=eyJkYXRhc2V0Q2F0ZWdvcnkiOiJmb3Jlc3RDaGFuZ2UiLCJtZW51U2VjdGlvbiI6IiJ9

(or go to Google search and type in 'Global Forest Watch' and go to 'map')

2. Explore the website noting the layer category menu at the far left. Click on LAND COVER and notice that the switch is ON for TREE COVER.

3. Go to the ANALYSIS option and click it to open. Click on the map to activate it. If it does not work try going to a new search engine and searching for Global Forest Watch.

4. Using Australia as the first example, record the following data if available in a table in Results: COUNTRY's NAME; TOTAL TREE COVER (%); TREE COVER LOSS (millions of hectares MHa); TREE COVER GAIN (MHa); TOTAL TREE COVER LOSS (%); PLANTATIONS (MHa); and COMMENTS (comment on the pattern of the graph given for Tree Cover Loss).

5. Drag the map across to another country by holding down the left mouse. Go to INDONESIA and click on the name of the country and then the ANALYSE button. Record the data in the table of Results as before.

6. Repeat this data gathering for each of the following countries: BRAZIL, CHINA, NEW ZEALAND, UNITED STATES and another of your OWN CHOICE.

EXPERIMENT 22.2 continued

RESULTS:

Draw a graph of the gathered data (note other resources below may have to be used to complete the table about Plantations) e.g.

COUNTRY	TOTAL TREE COVER (%)	TREE COVER LOSS (MHa)	TREE COVER GAIN (MHa)	TOTAL TREE COVER LOSS (%)	PLANTATIONS %	COMMENTS
Australia						
Indonesia						
Brazil						
China						
New Zealand						
United States						

Also see: https://farm1.staticflickr.com/300/31552050673_1167977beb.jpg
(map showing changes in forest plantation growth by country)
http://www.fao.org/docrep/w4345e/w4345e03.htm
(scroll down to details about plantations Tables 2 & 3)

VIDEOS:

1. Go on a mini-excursion to Puerto Maldonado, a logging and gold-mining town in the Peruvian Amazonia, with the author and his wife in 2011 and travel down the Rio Madre de Dios (River of the Mother of God) and into the rainforest at:

   https://www.youtube.com/watch?v=VhJ7Ve1FbL0 (38 minutes)

2. State of the world's forests with special reference to the Amazon from GRID-Arendal - a Norwegian foundation working closely with the United Nations Environment organisation.

   http://www.grida.no/resources/8444 (3 minutes)

QUESTIONS:

1. What factors must be kept in mind when looking at the total forest cover of any country?

2. Which of the selected countries has the highest (a) forest lost and (b) forest gained (c) amounts of plantation?

3. Convert the plantation values (MHa) in Question 10 (c) to percentages of total forests and rank the listed countries in order of plantation percentages.

4. Looking at the graphs of tree cover loss (from the Comments), which of the listed counties has the biggest problem with deforestation?

5. Not all of the values for deforestation are due to timber being logged as a resource. List other reasons for deforestation (some research needed).

EXPERIMENT 22.2 continued

6. How does deforestation affect climate change?

7. "Plantations are not true forests." Discuss this statement from an ecological viewpoint.

8. Are there any problems associated with timber plantations?
   (see: http://www.forestnetwork.net/Docs/Gippy_h2o.htm#plantations)

9. What timbers are used worldwide in plantations and as fast-growing trees for re-forestation? Indicate some case studies where these timbers are being used.

10. What are some strategies in restoring the world's forests and using timber as a sustainable ecosystem resource?

    See also:
    http://www.global-economic-symposium.org/knowledgebase/the-global-environment/protecting-and-restoring-the-world2019s-forests/figure_overlay
    Changes in world biomass

    http://www.fao.org/docrep/003/y0900e/y0900e05.htm
    (scroll down to Figure 8)

    http://www.forestnetwork.net/Docs/AlternativeVision.htm
    Future alternatives to logging

CONCLUSION:

Write a short report on the current state of the world's forest resources, causes of depletion and strategies being used to restore native forests and use timber as a sustainable resource. Include any diagrams, maps or photos as required giving appropriate references and acknowledgements.

RESEARCH (Optional)

Use the Internet to find out how renewable timbers can replace current building and other materials which use non-renewable resources as their raw material.

EXPERIMENT 22.3 Time: one to two lessons

SUSTAINABILITY of SURFACE WATER as an ECOSYSTEM RESOURCE

AIM: To examine the potential sustainability of surface water as an ecosystem resource.

MATERIALS: Computers or tablets with Internet connections, graphing facilities – electronic or paper.

BACKGROUND:

Surface freshwater is limited and unequally distributed in the world. Most countries rely on local rainfall or rivers from other countries which do have adequate rainfall. With most countries there is a supply/demand problem with increased need for agriculture, industry and drinking water putting considerable stress on river systems. Global warming models suggest that some countries will have diminishing water supplies that become even more scarce and so a sustainable surface water supply has become a major priority with governments. Sustainable water systems involve the provision of adequate water quantity and appropriate water quality for all local needs, without compromising any future ability to provide this capacity and quality. As rainfall will vary in the future, water sustainability will depend upon storage systems and general conservation with better techniques in agriculture, waste treatment, industrial and domestic use. This may also include systems where the use of water has traditionally been required but can be replaced by water-less systems such as dry sewerage systems.

PROCEDURE:

1. As a general introduction, open the website (Site 1.) at:

    https://www.iwapublishing.com/news/sustainability-water-supply

    Read the information carefully and use it to answers some of the questions later.

2. Open the UNESCO report on Nature-based Solutions (NBS) for Water 2018 at (Site 2):

    https://reliefweb.int/sites/reliefweb.int/files/resources/261424e.pdf

    It is a long and detailed report so specific attention is to be focussed on the PROLOGUE, pages 9 to 20 (Document pages 21 – 32) and Figure 1. (p.12), Figure 4. (p.15) and Figure 6. (p.18).

    If possible, open this and the next website together and toggle between both sites.

3. Open the Powerpoint presentation from the World Bank at (Site 3):

https://www.google.com/url?sa=t&rct=j&q=&esrc=s&source=web&cd=1&ved=2ahUKEwjT3PSF7d_fAhWMXrwKHf4oB7wQFjAAegQIChAC&url=http%3A%2F%2Fsiteresources.worldbank.org%2FEXTABOUTUS%2FResources%2FWater.ppt&usg=AOvVaw3OqTtP7Mbv6xIKG6CGxNx8

EXPERIMENT 22.3 continued

RESULTS:

Make any brief notes in point form from the general reading of the three sites.

QUESTIONS:

1. What are major sources of the World's freshwater and their relative percentages? (See Sites 1 & 3)

2. How are Australia and the United States generally assessed for water scarcity (Figure 1 Site 2 and Slide 5 Site 3 – Note: SEI index has red as critical)? Comment on the main regions of these countries affected.

3. What countries are regarded as uniformly very critical when it comes to water scarcity? (see references above especially Slide 5 Site 3.)

4. Comment on the use of groundwater (i.e. underground water – artesian and sub-artesian) as support for groundwater in irrigation and stock raising (Site 2, p 13 & 14. Note: 'Abstraction' refers to extraction of groundwater).

5. What is a major problem with contamination of surface water? (Site 2, pp. 14-17).

6. What will be one of the main problems with soil moisture due to future global warming? (Site 2, Figure 6, p.18).

7. What are some ways in which water can be conserved? (Site 1, Site 2, Figure 1.4, p.31, Table 1.2., p.32).

8. What will be some of the ecosystem services which will be used in improving water use and sustainability? (Site 2, Table 1.1., p. 30).

CONCLUSION:

Write a general assessment in several paragraphs concerning the availability of water in the future, the types of demands, both natural and man-made which will impact on available water sources and some of the possible ways by which water will be conserved and used as a sustainable resource. In your assessment use diagrams and data as required to support the content.

RESEARCH: (Optional)

1. Continue with the examination of website 3 (Slide 8 onward) and note the major current and future problems with developing countries such as those in Africa. What are the problems and what political tensions are likely to develop over water supply?

or

2. Use Site 2 to examine more closely the problems with ground (sub-surface) water. Some specific Internet research may be required with specific reference to Australia or the United States.

EXPERIMENT 22.4   Time: one to two lessons

CASE STUDIES - SURFACE WATER as a SUSTAINABLE RESOURCE

AIM: To examine the potential sustainability of surface water as an ecosystem resource with specific reference to either (a) the Murray-Darling River System of Australia or (b) the Colorado River System of the United States.

MATERIALS: Computers or tablets with Internet connections, graphing facilities – electronic or paper.

BACKGROUND:

The interior of Australia and the south western part of the United States contain some of the driest land in the world. Surface water in these regions is under extreme threat with global warming and extended agriculture and stock grazing. Sustainability of surface water in the two great river systems of these countries, namely the Murray-Darling River system of eastern Australia and the Colorado River System of the south western United States, is the subject of this activity. One of these case studies or both can be examined.

PROCEDURE:

PART A: Sustainability of the Murray-Darling system.

PROCEDURE:

1. Open the website prepared by the Commonwealth Scientific and Industrial Research Organisation (CSIRO) at:

    https://publications.csiro.au/rpr/download?pid=legacy:530&dsid=DS1

2. Read the report carefully with special reference to pages 28 to 46 and Appendices A and B.

RESULTS:

Make brief notes in point form of any useful information and references which would be useful in answering the questions below.

QUESTIONS:

1. Compare the current available surface water resource with its use (Key Findings 1 & 2).

2. What will be the most likely scenario of water availability in the future? (Key Findings 3).

3. What will be the effect of climate change on the water availability of this system? (Key Findings (7, 8 & 9).

4. What is the current surface water use in this system? (p.32 and charts p.33).

EXPERIMENT 22.4 continued

5. What is the probable impact of climate change on surface water availability? (pp. 34-38).

    For further information about future plans for restoration of the Murray-Darling system see also

    https://ac.els-cdn.com/S2210784315000790/1-s2.0-S2210784315000790-main.pdf?_tid=754f1479-6b28-40b4-8511-46c384e3b0de&acdnat=1547016042_ae3592b3a66f55dd7fd49416284982b6
    (a formal paper on sustainability of the system)

    and

    https://www.mdba.gov.au/managing-water/environmental-water/basin-wide-environmental-watering-strategy
    (information from the Australian government's Murray-Darling Basin Authority with further links at the bottom of the page.)

6. List and explain some of the strategies which will be used to assist in the sustainability of this system as a water resource.

CONCLUSION:

Write a brief report in paragraph form explaining the threat to the water sustainability of the Murray-Darling system in eastern Australia giving details of trends in availability, causes of water shortages, the possible effects of global warming and what attempts are being made to rectify the situation.

PART B: Sustainability of the Colorado River.

PROCEDURE:

1. Open the website prepared by the National Water Institute of the United States at:

    http://nwri-usa.org/pdfs/2010ClarkePrizeLecture.pdf

2. Read the report with special reference to the drivers which change water availability, quantity and quality and the various graphs and figures in the article to get a general idea of water availability in the United States.

3. Open the websites at:
    https://pacinst.org/issues/sustainable-water-management-local-to-global/colorado-river/ (Site 1 - a brief introduction)

    and

    https://www.usbr.gov/watersmart/bsp/docs/finalreport/ColoradoRiver/CRBS_Executive_Summary_FINAL.pdf (Site 2 - more detailed information)

EXPERIMENT 22.4 continued

4. Read both articles carefully – Site 2 will be the most informative.

RESULTS:

Make brief notes in point form of any useful information and references which would be useful in answering the questions below.

QUESTIONS:

1. Referring to the graph on Site 1:

   a. What was significant about the years 1996 and 2003?
   b. What is the likely projection of these curves past 2008?
   c. What was the possible cause of such anomalies in 1996 and 2003?

2. Other than the Colorado's drainage basin, what other areas take water from it? (See Site 2, Figure 1).

3. What is the Law of the River? Is it a government regulation? Explain (Site 2, p.3).

4. Who receives the water from the Colorado River? (Site 2, p.4).

5. What are the four possible future scenarios suggested by river studies? (Site 2, pp. 6 & 7).

6. What are some of the suggestions proposed to redress some of the problems with the Colorado water supply and allow for its sustainable use? List these and explain (Site 2, pp. 11-16, especially Tables 2 which is on two pages).

CONCLUSION:

Write a brief report in paragraph form explaining the threat to the water sustainability of the Colorado River system in south western United States giving details of trends in availability, causes of water shortages, the possible effects of global warming and what attempts are being made to rectify the situation.

RESEARCH (Optional)

Use the Internet to find out about the current problems with these rives and how they are being solved.

EXPERIMENT 22.5   Time: one lesson

GEOTHERMAL ENERGY as a SUSTAINABLE RESOURCE

AIM: To examine the sustainability of geothermal energy.

MATERIALS: Computers or tablets with Internet connections, graphing facilities – electronic or paper.

BACKGROUND:

Geothermal energy comes from the natural heat of the Earth which can produce hot water or steam at a shallow depth which can be drilled and pumped to generate heat or electricity by steam-driven turbines. There are two main types of geothermal resources: convective hydrothermal resources, where the earth's heat is carried by natural hot water or steam to the surface; and hot dry rock (HDR) resources, where fluids must be injected from the surface to capture the heat and bring it back to the surface. Geothermal areas are categorised as low- and high-temperature fields, where high-temperature fields have temperatures over 180°C and are found around tectonic plate boundaries where volcanic activity is high. Low-temperature fields can hold a range of resources, held as heat in deep crystalline rocks or from water travelling through faults and fractures.

PROCEDURE:

1. As an introduction, open the website of the World Energy Council at:

    https://www.worldenergy.org/data/resources/resource/geothermal/?gclid=EAIaIQobChMIwfWOwLDg3wIVyquWCh19eA8qEAAYAiAAEgIHlvD_BwE (Site 1)

2. Read this brief introduction carefully noting the main geothermal regions and some of the amounts of energy used (Note: Mtoe is sometimes used as an energy unit. This stands for one million tonnes of oil equivalent. This is the amount of energy released by burning one tonne (metric ton) of crude oil. It is approximately 42 gigajoules or 11,630 kilowatt hours).

3. Open the following website which shows a graph (page 3) for an operational hydrothermal well:

    https://orkustofnun.is/gogn/unu-gtp-report/UNU-GTP-2003-01-01.pdf (Site 2)

4. Look at this graph, especially the units and scale of the graph and its shape. Note any details in the Results.

5. Open the interactive website at:

    https://energyeducation.ca/encyclopedia/Geothermal_energy (Site 3)

6. Make general notes about geothermal energy, the sources of the heat and the sustainability of this resource.

EXPERIMENT 22.5 continued

On this website, scroll down to DATA VISUALISATION and look at how this is presented. In the results, prepare a table showing: country; and energies as a percentage (by clicking on the parts of the pie graph or the dots at its right): e.g. geothermal; coal; natural gas; nuclear; hydro; solar/wind/other; and comments about the relative percentages of geothermal compared to the main energy source for that country.

To get data for each of the selected countries, go to the LOCATION box at left, delete the name there and a drop-down menu will appear. Select the country then note the data in the results table. Choose the countries of: the World (total); Australia; United States; New Zealand; Iceland; Sweden; Japan; China; Turkey; Indonesia; Russia; and another of your own choice.

7. Open the following technical paper at:

   https://www.geothermal-energy.org/pdf/IGAstandard/Poland/2001/a4.pdf   (Site 4)

8. Read the document carefully noting the factors which control the sustainability and renewability of geothermal energy.

RESULTS:

1. Make brief notes in point form of any useful information and references from all four sites which would be useful in answering the questions.

2. Construct a table showing geothermal energy compared to other energy sources in some selected countries. Some countries may have zero values for some energies.

COUNTRY	ENERGIES GIVEN as a PERCENTAGE						COMMENTS
	GEOTHERM.	COAL	NATURAL GAS	NUCLEAR	HYDRO	SOLAR/ WIND/ etc.	
WORLD							
AUSTRALIA							
UNITED STATES							
NEW ZEALAND							
ICELAND							
SWEDEN							
JAPAN							
CHINA							
TURKEY							
INDONESIA							
RUSSIA							
OTHER							

QUESTIONS:

1. What is the difference between the use of geothermal power and geothermal heat? How much of each were used in 2014-2015? (Site 1).

2. On this site and others, energy units are in often in PJ and Mtoe. What are these units? power is also given in GW. What is this unit? (extra research needed).

3. What is the difference between energy and power? (research).

EXPERIMENT 22.5 continued

4. Look at the graph at Site 2 which shows the depletion of flow rate from an operation hydrothermal well. Use the graph to find out when (year) the well will have zero flow.

5. From the table constructed of geothermal and other energies for selected countries:

    a. Comment on the world use of geothermal energy, giving its percentage use and rank compared to the other energies.
    b. rate the selected countries in order of their use of geothermal.
    c. Of those countries having high geothermal use, comment on their use of other renewable energies (some addition may be needed).

6. What are the factors which control the sustainability and renewability of geothermal energy?

CONCLUSION:

Write a brief report in paragraph form explaining:

a. why geothermal energy is considered a renewable and sustainable form of energy;
b. what are the necessary sources of geothermal power and the different types available;
c. What are the limitations of geothermal energy; and
d. What is its current use and potential for the future?

RESEARCH: (Optional)

1. Use the internet to find the location of major geothermal power stations.

2. Distinguish between hydrothermal and hot dry rock (HDR) power systems. Use a simple diagram to explain how each generates power from the Earth's heat.

EXPERIMENT 22.6   Time: one lesson

POROSITY AND PERMEABILITY

AIM: To measure and compare the porosity and permeability for some common sediments.

MATERIALS: Sand, silt & gravel, measuring beakers, measuring cylinders, large filter funnels, filter paper, clock or timer with second hand.

BACKGROUND:

Hydrologists (water scientists), Agriculturalists, Agronomists (soil scientists) need to know the porosity and permeability of soils and sedimentary rocks.

POROSITY is the amount of pore space (open space) between the grains of sediment. It is found by:

$$\text{POROSITY \%} = \frac{\text{Volume of pore space}}{\text{Total volume}} \times \frac{100}{1}$$

PERMEABILITY is the rate at which water moves through sediment and this depends upon several factors including the nature of the sediment, the cross-sectional area of the layer of rock and the height of water intake over outflow or the head of water.

PROCEDURE:

PART A: Porosity

1. Fill the measuring beaker to the 250ml level with sand or silt or gravel and check that the surface is level. This is the **total volume** of the soil.

2. Measure exactly 100ml of water in the measuring cylinder and slowly pour it into the beaker until the water level JUST reaches the soil surface. Record how much water was poured into the beaker. A second 100ml measurement may be needed with some soils so keep a record of the total amount of water poured into the beaker.

3. Determine the total amount of water poured into the beaker. As the water will fill the pore spaces, this amount is the volume of pore space.

4. Calculate the porosity as a percentage for this sediment.

5. Carefully empty this wet sediment into the appropriate waste bucket (one for EACH sediment) and repeat this method for the other two sediments.

EXPERIMENT 22.6 continued

PART B: Permeability

1. Place a small wad of torn filter paper into the bottom of the filter funnel.

2. Measure 100 ml of water in the measuring cylinder and transfer it to the measuring beaker.

3. Place the filter funnel into the top of the empty measuring cylinder and, starting from a set zero time, pour as much water as possible into the funnel (and maintain the level by constant pouring) timing how long it takes for all of the water to flow through the funnel. This control experiment will give the zero error due to the resistance of the wad of paper.

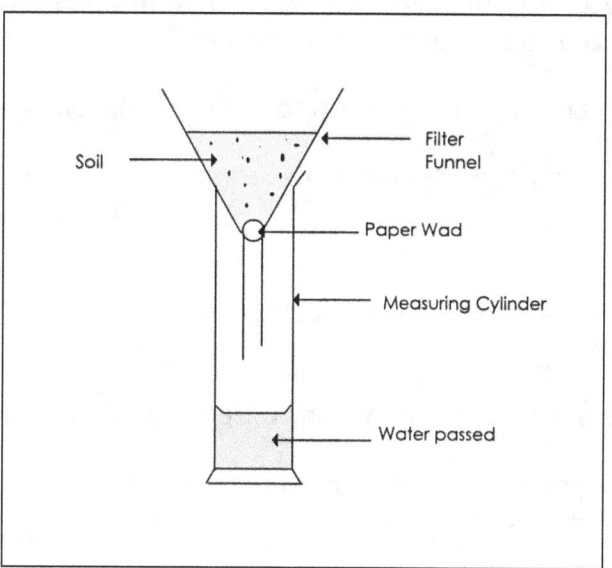

RESULTS:

Give a general description of what happened in both parts.
Construct tables such as:

PART A: Porosity

SEDIMENT	TOTAL VOLUME of SEDIMENT	VOLUME of WATER ADDED	POROSITY %
SILT			
SAND			
GRAVEL			

Calculations:

EXPERIMENT 22.6 continued

	SILT		SAND		GRAVEL	
	Dry	Wet	Dry	Wet	Dry	Wet
TIME						
VOLUME PASSED						

Calculations:

QUESTIONS:

1. Why are hydrologists and other people interested in the porosity and permeability of soil, sediment and rock?

2. Explain how a rock could be porous but not permeable? Why would this matter to exploration geologists looking for water or oil?

3. What is the purpose of the paper wad in the permeability experiment?

4. What is the purpose of the control experiment?

CONCLUSIONS:

1. List the sediments in order from best to least for (a)porosity and (b) permeability;

2. Are these lists the same? Explain any differences or similarities.

3. Which of the permeability methods (dry or wet) would be a more honest measure of a sediment's true permeability? Explain.

4. What could reduce the porosity of a sediment in nature?

5. Comment on any errors which could occur in the measurements.

RESEARCH (Optional)

How can rock have its porosity and/or permeability changed? How are these rock parameters useful in engineering of roads, dams, bridges and building construction?

Chapter 23: Renewable Energies

EXPERIMENT 23.1 Time: one lesson

HYDROELECTRICITY and WIND POWER

<u>AIM:</u> To demonstrate the production of electricity from the force of moving fluids.

<u>MATERIALS:</u> paper template, aluminium pie dish, scissors, small hollow tube (e.g. 2 cm of small drinking straw), blu-tack or other gum, tape or turbine/motor kit, portable fan, voltmeter, wires

<u>BACKGROUND:</u>

Hydropower and wind power have been used since ancient times to grind flour, pump water and perform other tasks. In the mid-1770s, the engineer Bernard Forest de Bélidor (French: 1798-1761) described both vertical- and horizontal-axis water wheel machines. In 1831, the development of the electrical generator by Michael Faraday (English: 1791-1867) enabled this new source of energy to be coupled with the old. In 1878 the world's first hydroelectric power scheme was developed at Cragside in Northumberland, England by William Armstrong (English: 1810-1900) and it powered a single arc lamp in Armstrong's art gallery. The first windmill used for the production of electric power was built in Scotland in July 1887 by James Blyth (Scottish: 1839-1906) and was used to charge simple batteries to power the lighting in his holiday cottage.

<u>PROCEDURE:</u>

<u>PART A:</u> Construction of a model water wheel.

1. Construct a simple paddle wheel by:

    a. Copying the following template onto paper then transferring it onto the base of a small aluminium pie dish.

    b. Trace the template onto the base of the dish or 'glue' it with water directly onto the dish.

    c. Cut along the solid lines of the template and then bend along the dotted lines to form a wheel with blades.

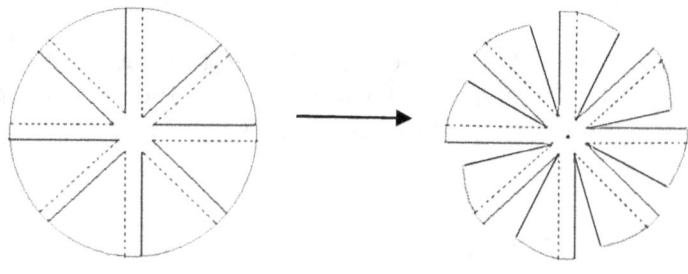

EXPERIMENT 23.1

2. Pass a small length of plastic tube (spaghetti tube) through a hole in the centre of the pie dish wheel and insert the other end in the shaft of the motor. Some experimentation may have to be used here to connect the wheel to the motor but one should be able to attach the shaft of the motor to a short piece of small drinking straw which comes with small cardboard cartons of juice. Blu-tack or gum or tape can be used to secure the tube to the centre of the wheel.

3. Hold the motor firmly in a clamp attached to a stand such that the wheel projects outwards:

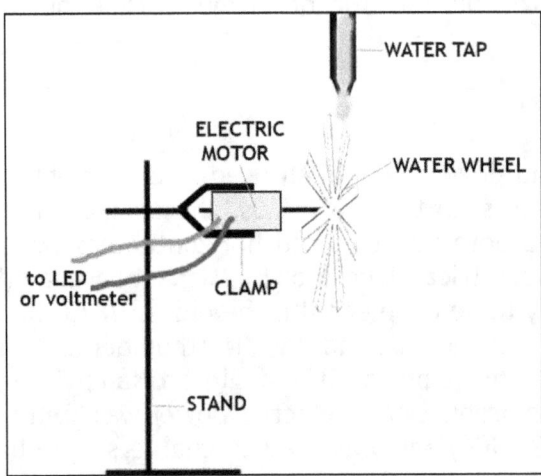

4. Alternatively, small kits of toy water turbines can be purchased for a cheap price. They often come with a geared wheel attached to a 9V motor and an LED.

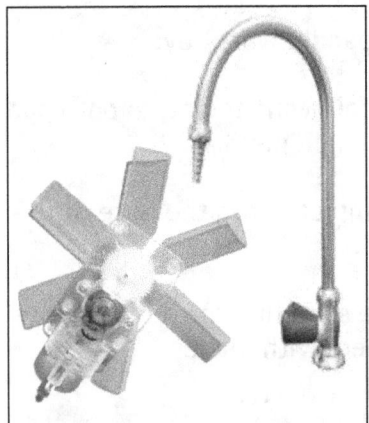

5. Position the turbine blades below the water tap and turn on the tap to give a slow stream. Notice the amount of light in the LED or the reading on the meter (preferred). Notice what happens when the water flow is increased.

EXPERIMENT 23.1

PART B: Wind power

Set up the apparatus or the commercial kit in a direct flow of air; a portable fan may be appropriate, and notice the change in the LED or the meter as the air flow is increased.

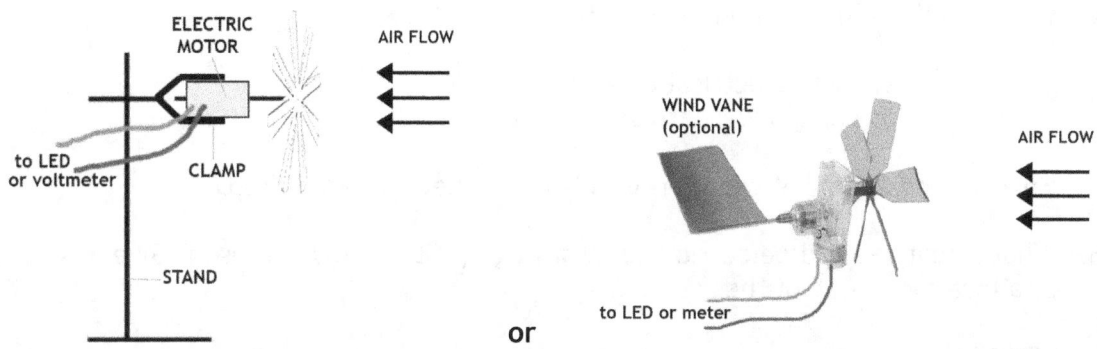

PART C: Real power requirements

1. Open the website previously use to obtain data for the world's energy production at:

    https://energyeducation.ca/encyclopedia/Geothermal_energy

2. Select different countries such as AUSTRALIA, UNITED STATES, CHINA, JAPAN and INDIA. Use the site to find the total percentages of hydroelectrical energy and wind energy used by these countries to copy and complete the table in the results.

RESULTS:

Make any observations about Parts A and B regarding the relationship between the strength of the fluid flow and the power developed.

Copy out and complete the graph for Part C:

COUNTRY	MAJOR SOURCE of ENERGY %	HYDRO ENERGY PRODUCED %	WIND, SOLAR & OTHER PRODUCED %
World	Oil 32% Coal 28%	2.4%	0.9%
Australia			
United States			
China			
Japan			
India			

EXPERIMENT 23.1 continued

QUESTIONS:

1. What is the relationship between fluid (water or air) and the amount of electricity generated?

2. Is it the amount of fluid or its flow rate which has the greatest effect?

3. How could this be regulated in hydroelectrical power stations?

4. What are the main requirements for:

    a. hydroelectrical power; and
    b. wind power farms?

5. How can the energy of wind farms be stored when the wind drops?

6. What is pumped hydroelectric energy storage (PHES)? Why is it used? Give an example of a location of one of these systems.

CONCLUSIONS:

1. Comment on the structure and processes required to make energy by these two methods.

2. Discuss the advantages and disadvantages of hydroelectricity and wind power generation.

3. From the table, what is the current importance of both of these forms of energy generation compared to other forms current used and is there any country which seems to be using these forms of power more than others and why?

4. What is the likelihood that these two forms of energy will play a more important role in energy production in the future?

    See:
    https://ourworldindata.org/renewables)

    https://www.nytimes.com/2017/02/09/business/energy-environment/wind-energy-renewable.html

RESEARCH: (Optional)

Use the Internet to find out the location of some of the world's largest wind farms and find out how they generate electricity.

Visit a wind turbine and see inside at:

    https://www.energy.gov/eere/amo/mining-industry-profile

| EXPERIMENT 23.2 | Time: one lesson |

SOLAR POWER

AIM: To demonstrate the production of electricity using solar voltaic cells.

MATERIALS: Photovoltaic cells (at least two or one per group), voltmeter, ammeter/milliammeter, wires, clips. Strong lamp if required.

BACKGROUND:

Solar panels are made up of many photovoltaic cells which convert sunlight into direct current electricity. There may be several solar panels connected to give a solar-power system. The photoelectric effect was first noted by the physicist, Edmund Bequerel (French: 1820 – 1891), who in 1839 found that certain materials would produce small amounts of electric current when exposed to light. In 1905, Albert Einstein (German: 1879 – 1955) described the nature of light and the photoelectric effect on which photovoltaic technology is based, for which he later won a Nobel prize in physics. The first photovoltaic module was built by Bell Laboratories in 1954 but was considered too expensive for everyday use but by the 1970s, photovoltaic technology gained recognition as a source of power. Solar cells are made of a thin wafer of semiconductor material such as silicon. This is specially treated to form an electric circuit, positive on one side and negative on the other. When light energy strikes the solar cell, electrons are knocked loose from the atoms in the semiconductor material forming an electrical current.

PROCEDURE:

PART A: Features of a voltaic cell

1. Connect a single voltaic cell to a voltmeter and measure its potential difference (voltage) in volts (V) making sure that the photocell is facing full Sun directly:

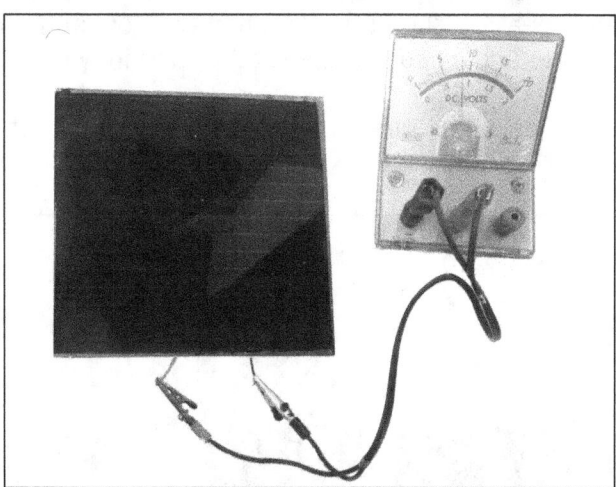

2. Change the voltmeter for an ammeter and measure the current (I) in amperes (A). If the current is small change the setting of the meter to a milliammeter and measure the current in milliamperes (mA).

EXPERIMENT 23.2 continued

3. Record both values in a table in the results and then calculate the power of the cell by multiplying the potential difference (volts) by the current (amperes):

    Power = potential difference x current

    P (watts) = V (volts) x I (amps.)

    e.g. if the potential difference (voltage) is 10 volts and the and the current is 65 mA (i.e. 0.65 amps), then the power of the cell is 10 x 0.65 = 6.5 watts.

4. Now connect two cells together in SERIES. This means that the negative terminal (black wire) of one cell is connected to the positive terminal (red wire) of the other. The other wires are then connected to the voltmeter.

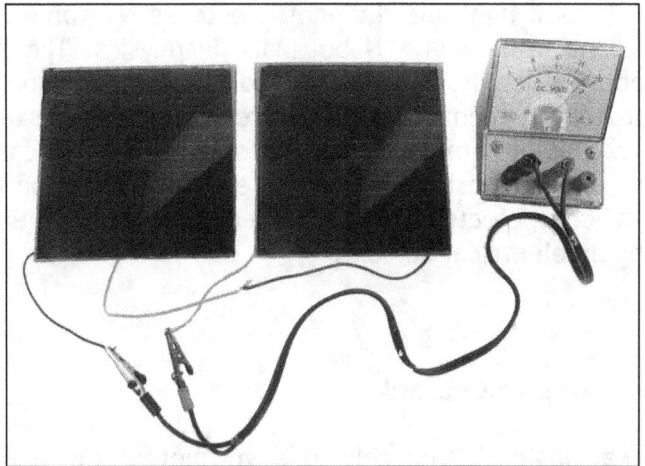

5. Measure the potential difference (voltage) and then replace the voltmeter with an ammeter and measure the current in amperes or milliamperes. Record both values in the Results table and calculate the total power in watts as before.

6. Reconnect the cells together in PARALLEL i.e. connect the red wires of both cells together and the black wires together and then connect them separately to the voltmeter:

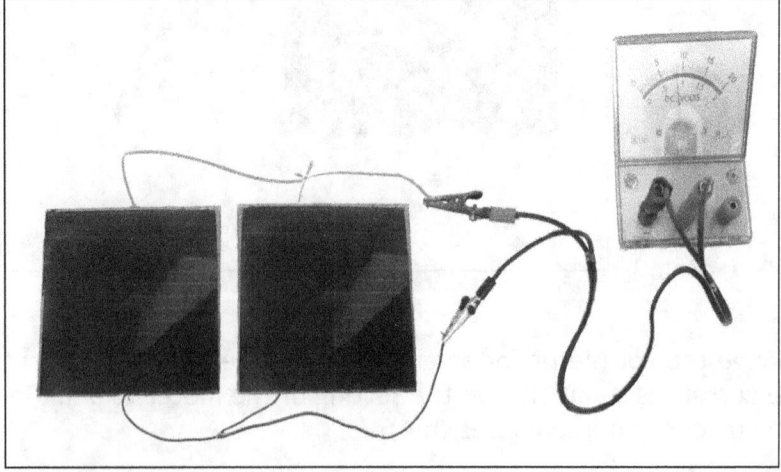

EXPERIMENT 23.2 continued

7. Measure the potential difference with the voltmeter and then replace it with the ammeter and measure the current. Calculate the power in watts as before and enter the values in the results table.

RESULTS:

Make any general comment about connecting the cells to the meters and to each other considering any errors which could occur.

Construct a table for the data measured e.g.

CONNECTIONS	SINGLE CELL	TWO CELLS in SERIES	TWO CELLS in PARALLEL	COMMENTS
VOLTAGE V (volts)				
CURRENT I (A or mA)				
POWER P (watts) = V x I				

QUESTIONS:

1. What factors would give a reduced value for the readings from the photocell?

2. What are the instrument errors of the (a) voltmeter and (b) ammeter? (remember the rules about analogue and digital scales).

3. What is the difference between a series and parallel connect in an electric circuit?

PART B: The efficiency of photovoltaic cells

Efficiency is the ratio as a percentage of the amount of energy given out by the solar cell compared to that put in by the Sun.

1. Watch the video at:

    https://www.youtube.com/watch?v=Am_BFu0EAuM

2. Use the value of the power (in watts) of the single cell calculated in PART A (3) by multiplying the value for its voltage and its current output (e.g. 10V at 0.65 A = 6.5W).

3. Calculate the energy input of the Sun's irradiation (flux) at this location by using the calculator at:

    https://www.cableizer.com/tools/solar_radiation/

EXPERIMENT 23.2 continued

Do this by:

    a. Opening the website and typing in the local city of town in the search box on the map at top right (e.g. Brisbane). This will give the LATITUDE (negative for Southern hemisphere and positive for Northern hemisphere e.g. -27 south) and the ALTITUDE above sea-level e.g. 15 metres a.s.l. Transfer these to the boxes at left.

    b. Type in the DAY of the YEAR as the number of days past January 1 e.g. if the date today was February 11, then the number would be 11 + 31 (i.e. date + days of January) e.g. 42.

    c. Type in the TIME of DAY using the 24-hour clock notation and as HH:MM e.g. if it is 7:30 am, then the time would be 07:30 (if it was 2:30pm it would be 14:30).

    d. Hit the CALCULATE button and scroll down to the value of the incoming solar energy in watts per square metre (W/m^2) e.g. 1115 W/m^2

4. Now calculate the efficiency of the solar panel by first measuring the surface area of the panel e.g. the panels shown in the photographs were 15 cm x 15 cm or 0.0225 square metres.

5. Next recall the value of the power (in watts) measure by the single cell (6.5 W). Divide this by the surface area (0.0225 m^2) to get the value of watts per square metres i.e 288W/m^2

6. Now calculate the efficiency of the cell by:

    Efficiency (η) = energy output/energy input   x   100/1   as a percentage

    e.g.   288 W/m^2 divided by 1115 W/m^2 x 100/1   = 26%

RESULTS:

Power output of the cell from class experiment =

Solar radiation(flux) input for the location at (latitude) with an altitude of (m) at the date of (   ) and time of (   ) =

CALCULATIONS:

Follow the steps and examples above to calculate the efficiency of the single cell used (or cells if different sized cells are used):

    Efficiency =   Energy output/Energy input   x   100/1   %

EXPERIMENT 23.2 continued

QUESTIONS:

1. What is the efficiency of the single cell(s) used?

2. What are some of the efficiencies of some common solar panels?
   (see: https://news.energysage.com/what-are-the-most-efficient-solar-panels-on-the-market/)

3. What are some of the factors which also determine the efficiency of a complete solar panel system?
   (see: https://solarcalculator.com.au/solar-panel-efficiency/)

CONCLUSIONS:

Generally, comment on the pros and cons of solar cells and how much power they can produce. Also comment on the method used and the errors involved.

1. What was the potential difference and current output of the single cell in full sunlight?

2. What was the efficiency of the single cell?

3. What was the effect on adding the cells together in (a) series and (b) parallel?

4. What is Ohm's Law? Does this apply to the connection of voltaic cells? Explain

EXERCISE: A typical solar panel system may be rated at 6.6 kW and have up to 22 individual panels of 300 W each. It may also be attached to one or more lithium ion battery packs for night use and an inverter for changing the direct current input to 240 v (110V in the US) alternating current.

Consider a typical home using only solar power as given above at meal time at night. The family may have switched on: a refrigerator (20 cf or cubic foot) and freezer (14cf) which are permanently on, a stove with two hotplates, an electric wall clock, 4 incandescent light bulbs at 100 W each, 1000 W microwave and a plasma TV.

Question: will the home solar panel system and battery allow all of this to happen?

see:
https://coolaustralia.org/wp-content/uploads/2013/12/Typical-power-ratings-for-appliances.pdf

RESEARCH: (Optional)

Use the Internet to find out how much energy and raw materials are needed to produce one 300 W panel. Are there any environmental problems?

EXPERIMENT 23.3   Time: one lesson

HYDROGEN GAS as a FUEL DERIVED from SOLAR POWER

AIM: To produce hydrogen gas using solar energy.

MATERIALS: Photovoltaic cells, electrolysis apparatus (Hoffman Apparatus) or lab. made (large beaker, metal electrodes, two test-tubes), water, dilute sulfuric acid, test-tubes, matches, taper or thin splints of wood, rubber gloves, 12 v power supply if solar cells cannot be used.

BACKGROUND:

Electrolysis of water is the decomposition of water into one part by volume of oxygen and two parts by volume of hydrogen gas due to the passage of an electric current. The reaction requires a potential difference of 1.23 volts to split the water which is usually slightly acidified with dilute sulfuric acid to act as a conductor and allow a small current to pass. Hydrogen fuel is a zero-emission fuel when burned with oxygen and can be used in electrochemical cells or internal combustion engines to power vehicles or electric devices. Since hydrogen gas has a very low density, it rises into the atmosphere and is rarely found in its pure form as the gas. When burnt hydrogen (H_2) reacts with oxygen (O_2) to form water (H_2O) and releases energy.

$$2H_2(g) + O_2(g) \rightarrow 2H_2O(g) + \text{energy}$$

PROCEDURE:

1. Wearing rubber gloves, fill the electrolysis apparatus with water to which has been added some dilute sulfuric acid to help as a conductor. It does not under go any chemical change. Take care when inverting the filled test-tubes and placing them, inverted, into the beaker:

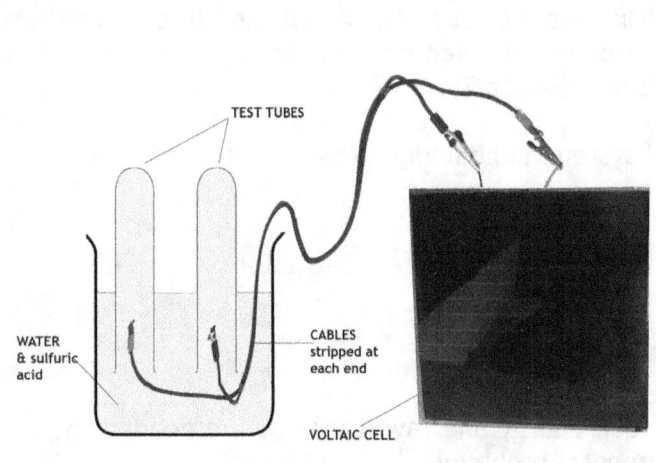
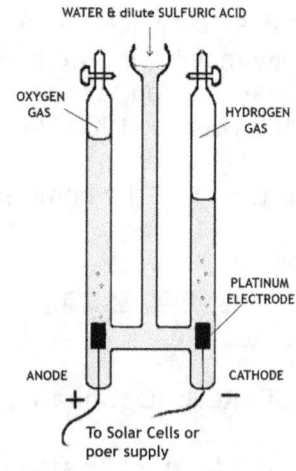

Experimental apparatus   **Hoffman's apparatus**

EXPERIMENT 23.3 continued

2. Place the apparatus with the voltaic cell in full sunlight and watch for any reaction. If the cell is of low power then this may take some time. Either connect several cells together in series or connect the wires to a low voltage power supply and adjust the amount of current (the cell shown in the photograph was 10V and 0.65A and gave satisfactory results).

3. After sufficient gas has been produced in each test tube, carefully remove the electrodes without disturbing the vertical test-tubes.

4. Wearing rubber gloves, carefully place the thumb over the mouth of the test-tube containing the hydrogen gas by inverting the hand.

5. Withdraw the test-tube, turn it the right way up and at the <u>same time</u> quickly remove the thumb and thrust a lighted match into the test-tube. Hydrogen gas should ignite with a loud pop if there is the right concentration of gas and air.

6. Repeat this test with the other test tube but thrust in a glowing taper to test for oxygen which will reignite the taper.

RESULTS:

Describe what happens when the electricity supply is connected to the apparatus. Explain the differences in volume of gas which is collected in each tube and why it is able to do so.

Also describe the tests for both gases, the result and their significance.

QUESTIONS:

1. Why must both test-tubes be completely filled with water?
2. Why was the sulfuric acid added?
3. Which test-tube will produce the hydrogen gas? Explain.
4. How can one observe that the water is splitting up by the electricity?
5. Why should one be very quick in performing the test with the lighted match or taper?

CONCLUSION:

1. What is produced by the electrolysis of water?
2. Was the use of the voltaic cells practical? If not how could this system be improved?
3. Is hydrogen gas as a fuel a good alternative to the current use of petroleum products? Why?
4. What waste product is produced by (a) electrolysis of water and (b) combustion of hydrogen gas fuel?

RESEARCH (Optional): Use the Internet to find out how hydrogen gas is current used in powering vehicles.

EXPERIMENT 23.4 Time: one lesson with extension
EXPERIMENT or DEMONSTRATION

MANUFACTURE of ETHANOL as a FUEL

AIM: To prepare alcohol (ethanol) from the fermentation of yeast and to distil the yeast mixture to obtain the alcohol formed by its anaerobic respiration.

MATERIALS: Yeast mixture (this may have to be left for several days to ferment further), double right-angled delivery tube, test-tubes, limewater, commercial distillation apparatus (as shown), water bath of heated water, watch glass or dish, matches.

BACKGROUND:

Sugar and any plant or animal product containing it can be fermented using natural yeasts which often grow on the outside of plants or it can be artificially added. Grains such as hops and corn, grapes, honey and even milk can be fermented to make alcohol. A potentially useful biofuel, alcohol (as ethanol) can be made in large quantities from sugar cane and then distilled. The overall reaction of the fermentation is:

$$C_6H_{12}O_6 \rightarrow 2\ C_2H_5OH + 2\ CO_2$$

PROCEDURE:

1. Prepare a several yeast mixtures for its fermentations several days before the experiment by dissolving about 20.0 g of sugar in 100 mL of tap water. To this was added a sachet (about 7.0 g) of yeast and the mixture was microwaved for 15 seconds in a 1.65kW microwave oven, less if it is more powerful at full power in order to fully activate the yeast. Alternatively, warm water at about 43 °C should give the recommended temperature range for activation.

2. A double right-angled glass delivery tube is passed through a cork and the end inserted into a clear solution of limewater. Limewater turns cloudy with carbon dioxide gas. A separate test-tube open to the air can be used as a control.

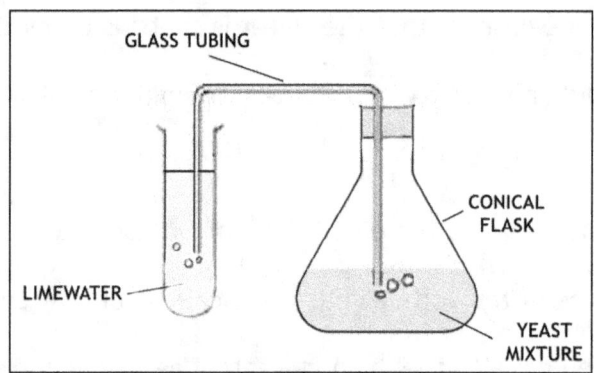

3. Observe the conical flask for any signs of fermentation then place it in a secure, dark area which is warm. Observe any changes after about one or two days.

EXPERIMENT 23.4 continued

4. When the yeast has fermented, set up the distillation as shown below. The boiling point of ethanol is 78°C and it is inflammible so all seals must be tight.

5. Maintain the temperature using boiling water in the water bath. This will ensure that the temperature in the boiling flask containing the mixture does not allow its water content to boil, only the alcohol should boil off.

6. When a few drops of alcohol have been produced (collected on the watch glass), stop heating, remove the watch glass to some distance and attempt to ignite the alcohol in a darkened room with a lighted match.

7. Allow the apparatus to cool and then wash it out thoroughly and put it away.

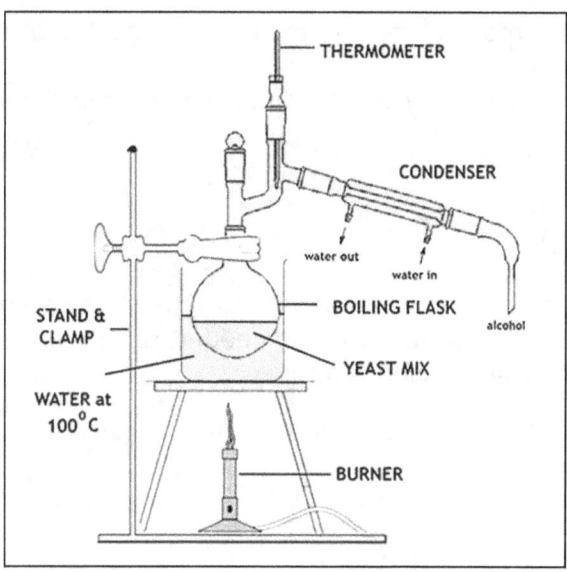

RESULTS:

Describe the results of the experiment noting the results of fermentation and how the distillation apparatus worked and if any liquid was produced.

Describe the result in attempting to ignite the liquid produced.

QUESTIONS:

1. What is yeast?

2. Why was the mixture microwaved or mixed in warm water?

3. What are some of the signs of fermentation occurring?

4. What was the purpose of the limewater? What did it show about fermentation?

5. How was condensate which came out of the distillation apparatus tested? What was the result?

EXPERIMENT 23.4 continued

6. What was the purpose of the thermometer at the top of the apparatus?

7. What is a condenser? Why was it important to have the water flow as indicated in the diagram?

CONCLUSIONS:

1. What is fermentation?

2. Was any alcohol produced? Did it ignite?

3. What was the original source of the alcohol?

4. Why was a water bath used and not direct heating with the burner?

5. What was the purpose of the thermometer at the top of the apparatus?

6. What is a condenser? Why was it important to have the water flow as indicated in the diagram?

RESEARCH: (Optional)

Use the internet to find out how ethanol is produced commercially and how it is currently used as a fuel. Mention the raw materials needed to produce the sugar and the advantages and disadvantages of using ethanol as a fuel.

Chapter 24: The Earth in Motion

EXPERIMENT 24.1 Time: one lesson

HOOKE'S LAW

AIM: To verify Hooke's Law that within the elastic limit of a material (such as springs, rock etc.), stress (applied force/area) is proportional to strain (distortion increase)

MATERIALS: Retort stand and clamp, ruler graduated in millimetres, helical spring, slotted masses and carrier, small bench (or G) clamps.

BACKGROUND:

Hooke's law is a principle of physics that states that the stress (force/unit area) needed to extend or compress a material such as a spring or rock by some distance X is linearly proportional to the strain (or increase in deformation). This will only occur within the elastic limit of the material, that is that limit when it will return to its original length or shape when the applied force has been removed.

In this experiment, the coiled spring represents the material to be stressed by adding weights to its end. The strain placed upon this spring can be measured in terms of the length that the spring extends as the weights are added. So, for the spring, Force applied (F) = kX, where k is a constant factor characteristic of the spring (the Spring Constant or its stiffness), and X is small compared to the total possible deformation of the spring.

This law is named after 17th-century British physicist) Robert Hooke (1635 – 1703) and it has been found also to be useful in rock mechanics when discussing the forces which may be applied to rocks causing them distort.

PROCEDURE:

1. Set up the apparatus as shown below:

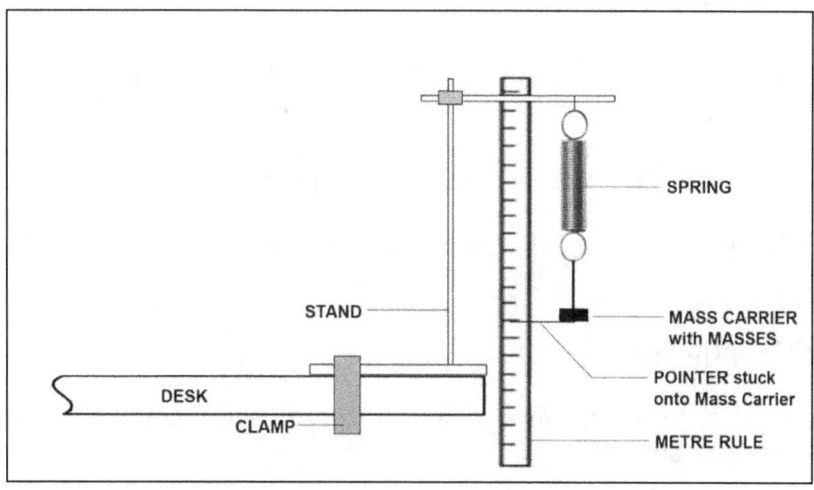

EXPERIMENT 24.1 continued

2. Record the measurement on the metre rule with just the empty mass carrier and its mass (usually 50 grams).

3. Add successive values of 50 g (0.05 kg) masses to the mass carrier and record the new measurement on the metre rule for the extended length of the spring in a table.

4. Repeat this addition of mass up to about 400 g (i.e. 0.400 kg) total, measuring the length from the starting position on the metre rule each time.

5. Calculate the force applied by the addition of the weights by multiplying the value in <u>kilograms</u> by 9.8 (m/sec^2) to give the force units in newtons (N).

6. Draw up a table showing the values for mass (g), the value for the force applied (kg) and the length of spring from the starting position (in mm).

7. Graph the values for the force applied on the horizontal axis and those for the length of the spring on the vertical axis.

RESULTS:

Redraw the diagram of the apparatus, construct the table as suggested and draw the graph accurately to a suitable size.

QUESTIONS:

1. Why is Hooke's Law valuable in understanding in how rocks fold or break along faults?

2. What is the difference between ductile and brittle structure? Which apply to faults and folds?

CONCLUSIONS:

1. What is the shape of the graph?

2. What does this shape suggest about the relationship between the length of the spring and the force applied to the spring?

3. Is this consistent with Hooke's Law? Explain.

4. What is the value of the Spring Constant (k)?

RESEARCH (Optional)

Find out about the life and work of Robert Hooke. Why would knowledge of Hooke's Law be useful in geology?

EXPERIMENT 24.2   Time: one lesson or demonstration

DEFORMATION of ROCK and EFFECTS of TEMPERATURE

AIM: To show the effects of temperature on deformation of a plastic material.

MATERIALS: Retort stand and clamp, ruler graduated in millimetres, Plasticine, slotted masses and carrier, small bench (or G) clamps, thermometers (say 0-100^0), clocks or watches with second hands.

BACKGROUND:

Some materials are plastic i.e. they do not obey Hooke's Law; when they are deformed even slightly, they stay deformed and if stressed too much or suddenly, may break.

PROCEDURE:

Note: this is the same apparatus as the previous experiment on Hooke's Law except that the elastic metal spring has been replaced by soft, plasticine.

1. Set up the apparatus as shown below with a long length of rolled plasticine fastened top and bottom by clear tape:

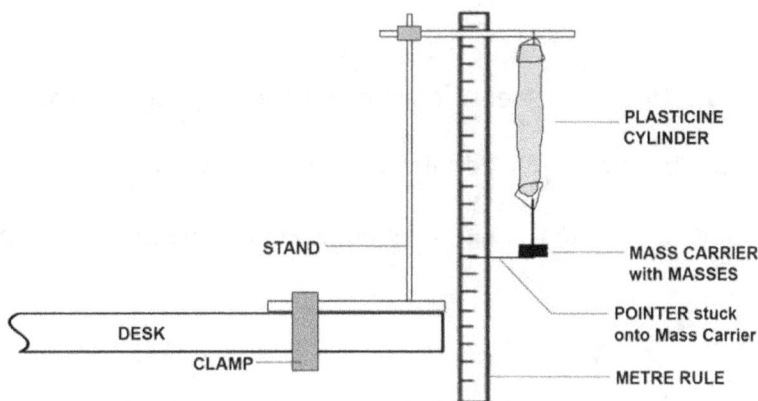

2. Take equal amounts of plasticine so that four equal cylinders can be rolled to about 6-8 cm in length and about 1 to 2 cm in cross-section.

3. Place a cylinder into a small plastic bag and immerse it into ice water (so that the bag keeps it dry) for two minutes.

4. Measure the temperature of the water.

5. Quickly remove it from the water and bag and, with minimal handling, attach it to the stand with tape. Attach the mass carrier to its base also with tape.

6. Note the length of the cylinder with the mass carrier attached using the pointer.

EXPERIMENT 24.2 continued

7. Add successive values of 50 g (0.05 kg) masses to the mass carrier, timing the experiment until the cylinder breaks. Record the minimal mass required to do this and the length of the cylinder. Also describe the deformation which occurred just before it breaks.

8. Repeat steps 3 to 7 using a cylinder which has been placed in ordinary water from the tap (faucet).

9. Repeat steps 3 to 7 for a cylinder placed in half tap water and half very warm water, being sure to measure the temperature of the water.

10. Repeat steps 3 to 7 using very warm water, recording its temperature.

RESULTS:

Redraw the diagram of the apparatus.

Draw up a table showing the: temperature of the water; the length at the start (50g mass carrier); total mass added for each step; the corresponding length of the cylinder; and the time and mass required to break the cylinder.

Describe any deformation fully for each temperature.

QUESTIONS:

1. Why isn't plasticine considered elastic and obeying Hooke's Law?

2. Why was the plasticine placed into bags before emersion in water?

3. Why is minimal handling required when attaching the plasticine to the stand and mass carrier?

CONCLUSIONS:

1. What is the relationship between temperature and the mass require to break the cylinder?

2. What is the relationship between the temperature and the time taken to break the cylinder?

3. Describe how the cylinder broke for each of the temperatures.

RESEARCH (Optional)

Other than external force and temperature, what other factors may be involved in the deformation of rock such as sedimentary rocks?

EXPERIMENT 24.3   Time: one lesson or demonstration

COMPRESSIONAL STRUCTURES

AIM: To show the effects of compression in the deformation of rock layers.

MATERIALS: Small, transparent rectangular plastic boxes, flour (or corn starch), fine sand, wooden rectangular block.

BACKGROUND:

When rock layers are compressed by large, regional Earth forces, they undergo deformation producing a range of structures such as anticlines, synclines, overfolds and sometimes shear faults.

PROCEDURE:

1. Place the wooden block vertically into the rectangular box at one of its ends so that it will act as a piston

2. Spread a layer of fine sand on the bottom of the box and smooth it out so that the top is horizontal.

3. Cover this with a layer of fine flour so that it makes a thin, horizontal layer over the sand.

4. Add another layer of fine sand on top of this and smooth to horizontal.

5. Add another thin, horizontal layer of flour on top of this sand layer.

6. Add the last layer of sand on top. The final apparatus should look like this:

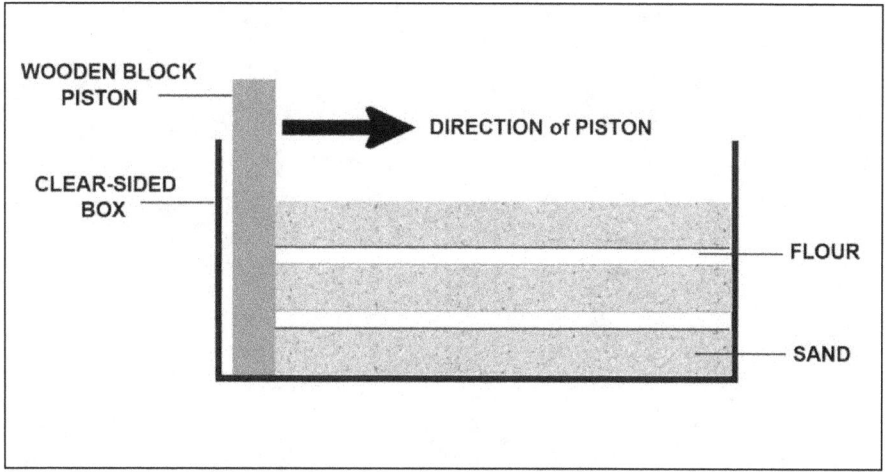

EXPERIMENT 24.3 continued

7. Move the piston <u>slowly</u> noting the effects continuously until extreme folding occurs and the limbs (sides) of the folds are over each other. Sketch these folds and give them an appropriate name (sketch 3).

8. On sketch 3, use a ruler and pencil to mark possible fault lines.

RESULTS:

Draw sketches 1, 2 and 3, labelling the types of folds and any faults seen in the model.

QUESTIONS:

1. Why is it important to move the piston slowly?

2. What is the purpose of the flour?

CONCLUSIONS:

1. Comment on the use of this sandbox model in showing folding and faulting in rock.

2. Relate the extent of the compression to the types of folding produced. Where in nature does this compression come from?

3. How are such models useful in the understanding of a dynamic Earth?

RESEARCH (Optional)

Use the Internet to find photographs of the types of folding and faulting seen in the model and then draw and label simple sketches of these.

Chapter 25: Volcanoes

EXPERIMENT 25.1   Time: one lesson

THE SHAPE of a VOLCANO

AIM: To draw a topographical cross-section of a volcano and deduce its possible eruptive type.

MATERIALS: Photocopies of the topographical map (below) graph paper, ruler and pencils

BACKGROUND:

Volcanoes often have distinctive shapes due to their eruptive type and the material produced by these eruptions, lava, ash or a combination of these.

PROCEDURE:

1. Look at the topographic map below which shows contour lines for a typical volcano of a particular type. Heights above sea level are given in metres and there is a deep crater lake at the summit.

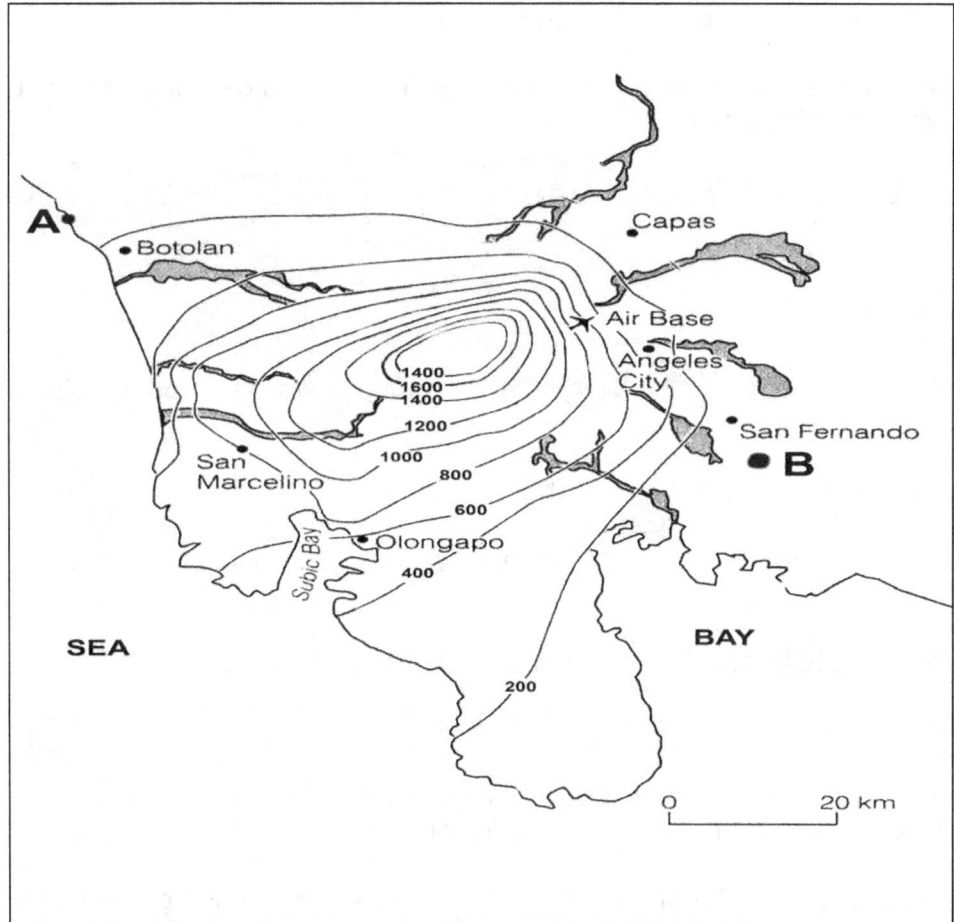

EXPERIMENT 25.1 continued

2. Draw a topographical cross-section between A and B on the map, choosing an appropriate scale for the vertical height. Students submitting an electronic report should do the cross-section on paper and then photograph or scan it. Remember, cross-sections are drawn by:

   a. Placing a straight edge of a piece of paper between A and B and marking where the contour lines cut the edge – mark the numbers also on the paper;

   b. Constructing the cross-section rectangle to a suitable vertical scale (e.g. 1 cm = 200 metres of height);

   c. Transferring the data from the paper's edge to the top of this rectangle and then drawing construction lines from each contour mark down to the appropriate height above sea level and marking this point with a faint dot;

   d. Joining the dots with a 'curve of best fit' to copy reality; and then

   e. Adding any appropriate labels or colour.

RESULTS:

Drawn topographical cross-section between A and B.

QUESTIONS:

1. What are the horizontal and vertical scales?

2. What is the approximate Vertical Exaggeration in numbers and what is its meaning? Why is this important?

(See: http://community.boredofstudies.org/23/geography/241383/how-calculate-vertical-exaggeration.html)

CONCLUSIONS:

1. How is this volcano most likely to be classified (as far as eruption material is concerned)?

2. Give reasons for your classification.

3. If the volcano has not erupted in many years, what would any new eruption likely to be in the first few hours?

4. What would be the main hazards to the nearby cities? Why?

RESEARCH (Optional)

1. Use the Internet to find out how volcanoes (especially those near centres of population) are monitored and predictions made about future eruptions.

2. What emergency action would be taken by the local population if there was an imminent threat of this volcano erupting.

EXPERIMENT 25.2

Time: one lesson

LOCATIONS of SOME MAJOR VOLCANOES

AIM: To plot the locations of some of the world's major volcanoes to see if there is a pattern in their locations and to note any relationships of eruptions to climate change.

MATERIALS: World map, table of major volcanoes and their locations (below).

BACKGROUND:

There are more than 1500 active volcanoes of several different types in the world today. There are many more which have not erupted in living history and so are considered extinct. Occasionally, new volcanoes occur below the sea and on land. This activity is an exercise in plotting the latitude and longitude of some of the world's major volcanoes to see if there is a pattern in the locations of volcanoes.

PROCEDURE:

PART A: Locations of some active volcanoes

On a map of the world (photocopied or printed), carefully plot the latitudes and longitudes for each volcano given in the table below by using the numbers for these locations. Remember to plot latitude first and then the longitude. Some locations are regions and so may be shaded for the entire length between the positions given:

NUMBER	NAME	RECENT MAJOR ERUPTIONS	PLACE or REGION	LATITUDE	LONGITUDE
1	Agung	1964 global ash	Bali, Indonesia	8S	115E
2	Asama	1783 local ash	Japan	36N	140E
3	Bezymienny		Kamchatka, Russia	55N	160E
4	Big Ben		Heard Island, Australian Antarctic Territory	55S	75E
5	Cosigüina	1835 global ash	Nicaragua	12N	34W
6	Cotopaxi		Ecuador	1S	78W
7	Misti		Peru	16S	71W
8	El Chichón	1982 global ash, SO_2	Mexico	17N	93W
9	Erebus		Antarctica	78S	165W
10	Mount Etna	1669 local lava and ash	Sicily, Italy	38N	15E
11	Fuego	1717 Local ash, lava	Guatemala	15N	90W
12	Fujiyama		Japan	35N	139E
13	Galunggung	1982 Local ash	Java, Indonesia	5S	105E
14	Izalco		El Salvador	13N	88W
15	Mount Katmai	1912 local ash	Alaska, USA	60N	155W
16	Kilauea	Many, lava	Hawaii, USA	20N	155W
17	Kilimanjaro		Tanzania, Africa	3S	37E
18	Krakatau	1883 ash	Indonesia	7S	105E
19	Laki	1784 SO_2	Iceland	64N	18W
20	Mt. Lamington	1951 local ash	Papua-New Guinea	9S	147E
21	Lengai	1960 local CO_3^{2-}	Tanzania, Africa	8S	30E
22	Mauna Loa		Hawaii, USA	20N	155W
23	Mayon	1814 lava	Luzon, Philippines	12N	125E
24	Nevado Del Ruiz	1985 lahars	Colombia	3N	76W
25	Ngauruhoe	1977 minor ash	North Island, New Zealand	38S	176E
26	Paricutin	1943 birth	Mexico	19N	108W
27	Mount Pelee	1902 local ash	Martinique	15N	61W
28	Pico Alto		Azores	27N	38W
29	Pinatubo	1991 ash	Philippines	18N	120E

EXPERIMENT 25.2 continued

Table continued:

NUMBER	NAME	RECENT MAJOR ERUPTIONS	PLACE or REGION	LATITUDE	LONGITUDE
30	Rabaul (several)		Papua New Guinea	4S	152E
31	Mount St. Helens	1980 ash USA	Washington USA	46N	122W
32	Nea Kameni		Santorini, Greece	38N	23E
33	La Soufriere	1902 local ash	St. Lucia	13N	61W
34	Stromboli		Italy	38N	12E
35	Surtsey	1967 lava	Off Iceland	63N	20W
36	Taal	1754 local ash	Luzon, Philippines	15N	121E
37	Mount Tambora	1815 ash	Indonesia	8S	118E
38	Tarawera		New Zealand	38S	176E
39	Mount Paektu		China	42N	130E
40	Vesuvius		Italy	41N	15E
41	White Island		New Zealand	38S	177E
42	Yasur		Vanuatu	20S	170E

RESULTS:

Plot the location of each location on the map using the numbers of each location. Students doing electronic reports will have to photograph or scan their completed map.

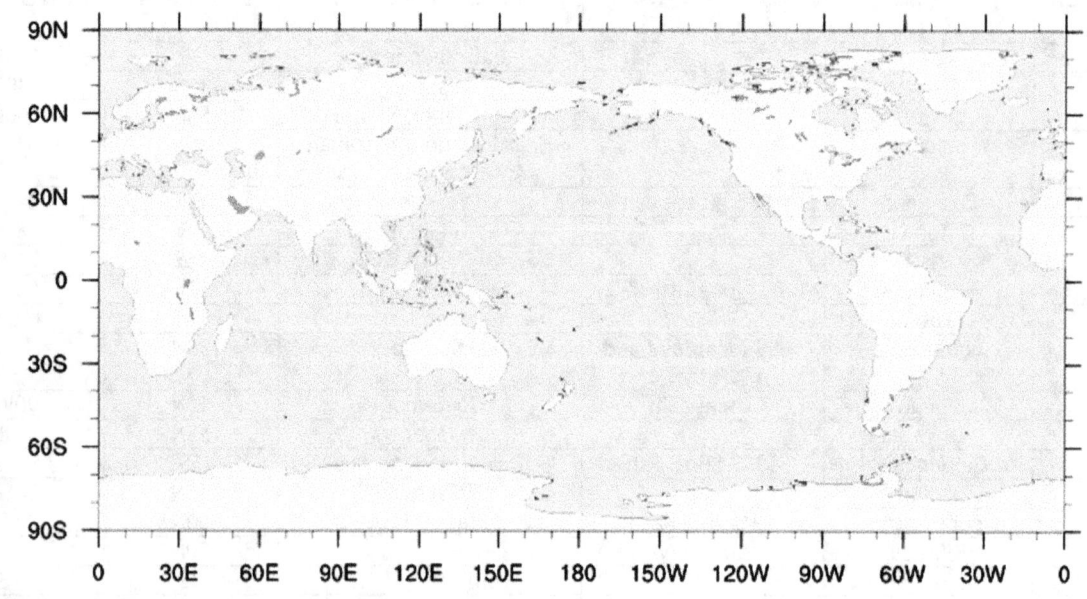

QUESTIONS:

1. How could this plot be improved?

2. What are the criteria which determine the activity state of a volcano?

EXPERIMENT 25.2 continued

PART B: Volcanoes and climate change

1. Examine the graph below of world temperature over recent time:

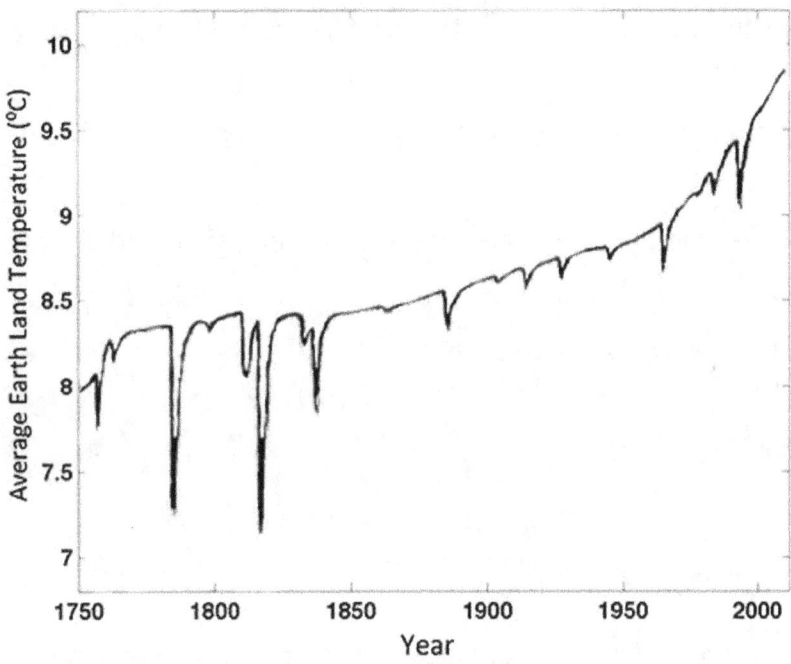

(Data from BerkeleyEarth.org)

2. Review the table of volcanic eruptions and match any of the major eruptions which may have affected world climate change.

RESULTS:

Describe any matches between volcanic eruption and possible climate change referring to the type of volcano and its scale of eruption.

QUESTIONS:

1. What is the general trend of this graph?

2. What does this graph generally show about climate change?

EXPERIMENT 25.2 continued

CONCLUSIONS:

1. Is there any pattern or logical groupings to the locations of the volcanoes?

2. Why do these volcanoes exist in these locations?

3. Account for the volcanoes in other locations where there might not be a grouping (e.g. Hawaii)?

4. How could a more accurate world pattern be obtained, especially in locations of isolated volcanoes?

5. Is there any relationship between some volcanic eruptions and climate change? Explain.

6. Could such a change occur in the future?

RESEARCH (Optional)

Use the Internet to:

1. locate currently erupting volcanoes

2. locate any places where a major volcano may erupt e.g. a region with a large amount of heat which could cause the eruption of a new or existing dormant volcano.

Chapter 26: Earthquakes

EXPERIMENT 26.1 Time: one or two lessons

EARTHQUAKE EPICENTRES and MAGNITUDES

AIM: To locate epicentres of earthquakes using triangulation from seismic waves.

MATERIALS: Computer or tablet with Internet, ruler with millimetre scale, calculator paper

BACKGROUND:

The location of an earthquake is usually given as its epicentre, the place on the surface directly above the source of the earthquake which is called its focus. Epicentres are found by triangulation from at least three seismic stations using sensitive seismometers and the pattern which they produce are called seismograms. Of the three types of earthquake waves (P, S and L), P and S waves travel through the Earth, with the S waves being slower than the P waves. The P waves will arrive first followed by the S waves depending upon how far away the epicentre is from the seismometer detecting the waves. Knowing the speed of each wave, the distance of the epicentre from the seismometer can be calculated. Triangulating at least three values from three widely separated seismometers, the location of the epicentre can be found.

PROCEDURE:

PART A: Locating the epicentre.

1. Examine the following three seismograms which were recorded at three seismic stations in Indonesia in 2018:

   SEISMOGRAM 1: Makasser Station:

   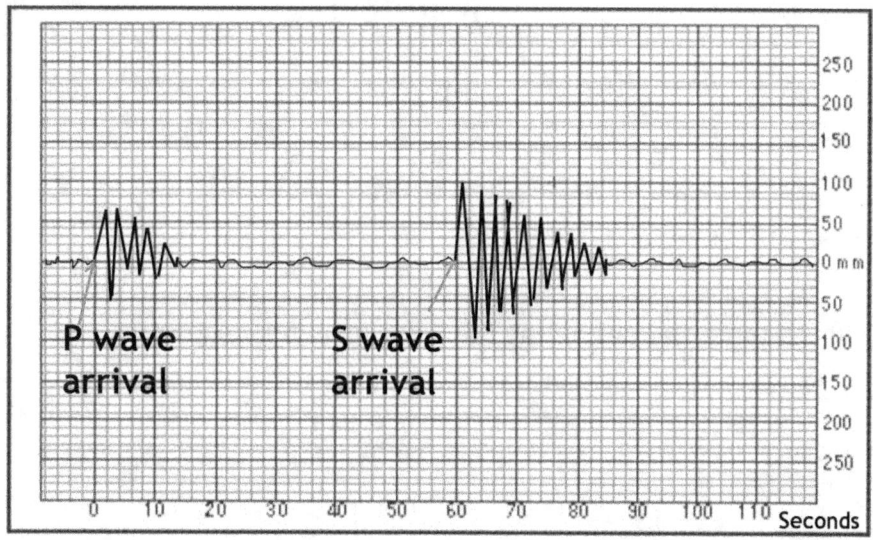

EXPERIMENT 26.1 continued

SEISMOGRAM 2: Palangkaraya Station:

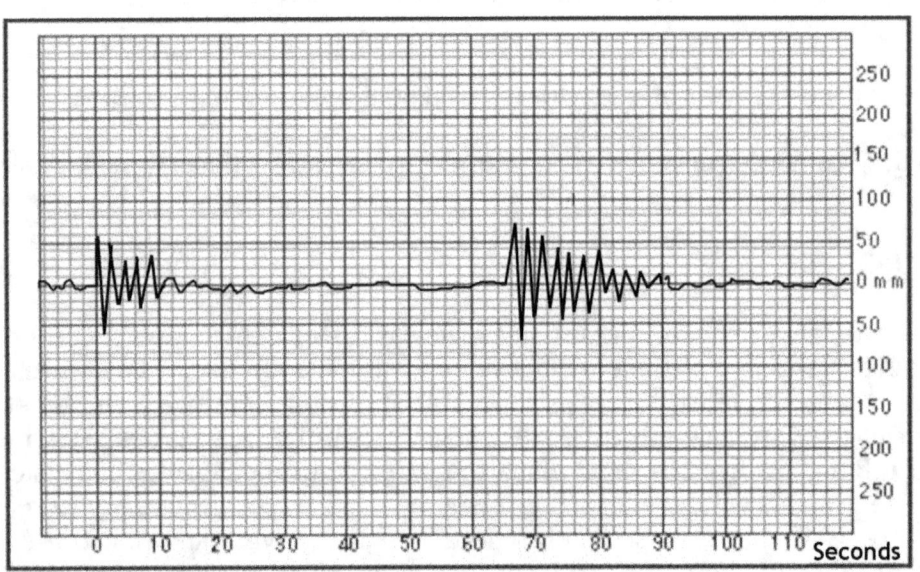

SEISMOGRAM 3: Tarakan Station:

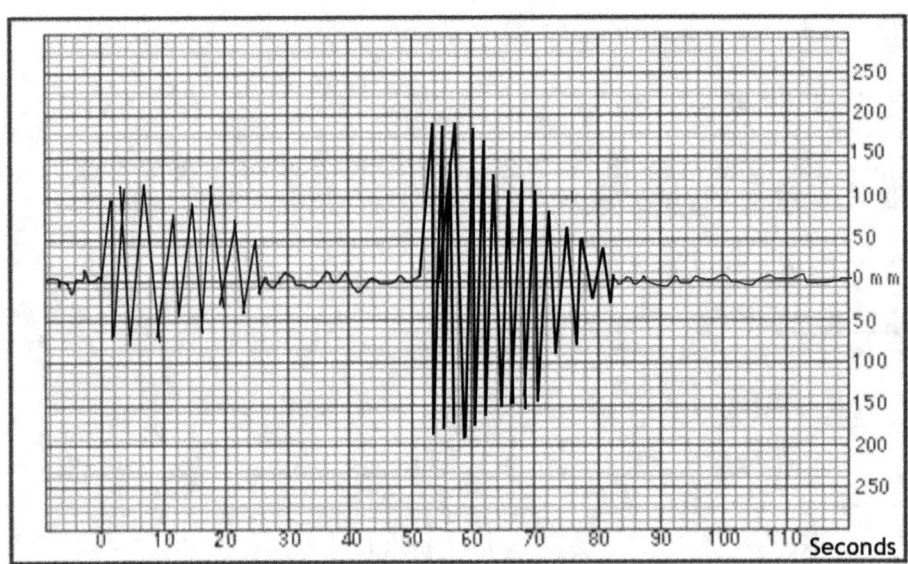

2. Locate these three stations on the following map of Indonesia:

EXPERIMENT 26.1 continued

MAP 1: Indonesian area:

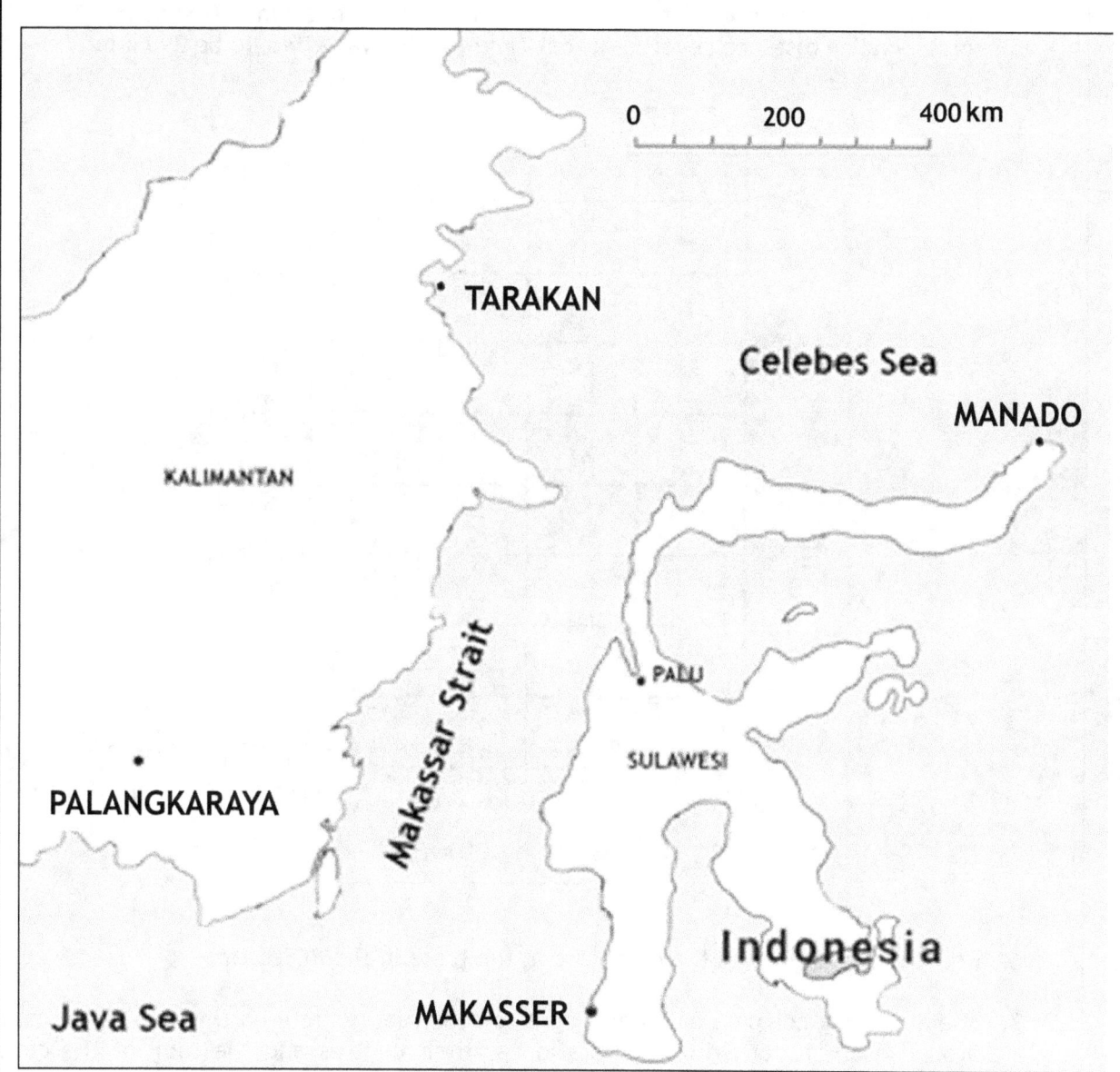

3. Print or photocopy this map on plain paper.

4. Measure the distance between the arrival of the P and S waves on each seismogram and record these distances in a table in the RESULTS e.g.

STATION	P - S TRAVEL TIMES (sec.)	DISTANCE from EPICENTRE (km)
Makasser		
Palangkaraya		
Tarakan		

EXPERIMENT 26.1 continued

5. Use the following graph, which shows the distance for the time differences between the arrival of the P and S waves, to find the actual distances between each station and the epicentre e.g. if the P-s travel time difference was 40 seconds, then the corresponding distance covered in that time for the waves would be 390 km:

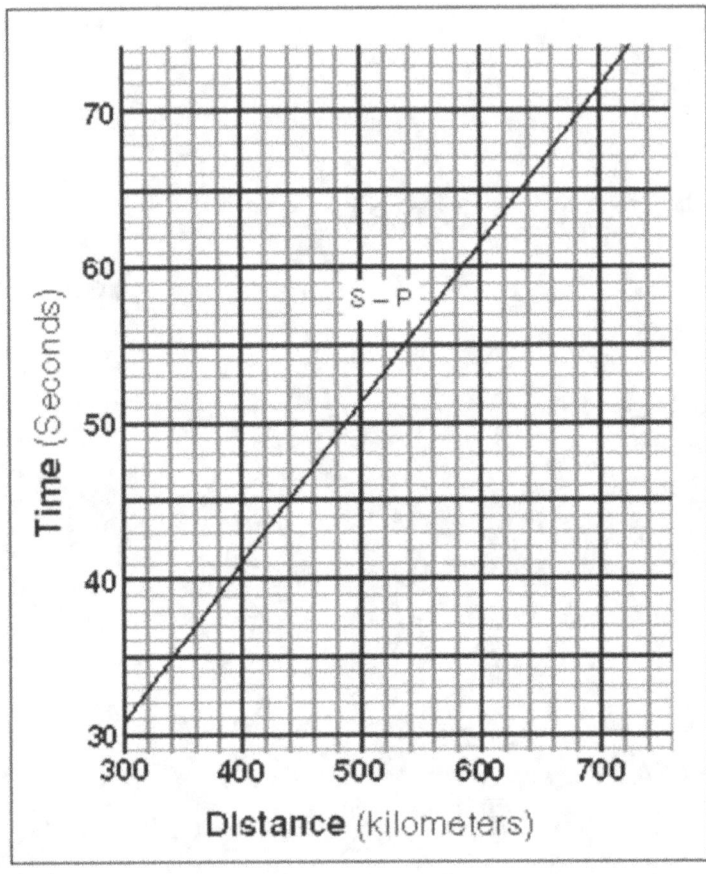

6. Record this distance to the epicentre in the table in the RESULTS:

7. Use a drawing compass or other means to construct a circle on the map, draw circles around each station with each station as their centres and the radii of the circles being the separate distances (in kilometres scaled to the map) to the epicentre.

8. Where the three circles cross should be the epicentre of the earthquake. If they overlap slightly, they will give a Triangle of Error. The epicentre will be in the centre of this triangle. Describe the location of the epicentre using scaled distance and magnetic compass directions e.g. 20 km SE of Tarakan.

EXPERIMENT 26.1 continued

RESULTS:

Record the data in a copy of the following table:

STATION	P - S TRAVEL TIMES (sec.)	DISTANCE from EPICENTRE (km)
Makasser		
Palangkaraya		
Tarakan		

QUESTIONS:

1. How are the P and S waves detected at each of the stations?

2. Why does the P wave arrive before the S wave?

3. Comment on the need for accuracy in making measurements from seismograms.

4. How could this accuracy be improved?

PART B: Finding the magnitude of the earthquake.

1. Measure the amplitude of each seismogram by measuring the distance from the zero line to the maximum height of the wave e.g. in seismogram 1, the amplitude is 240 millimetres (mm).

2. Record this height for each of the three seismograms and also record the distance of each station from the epicentre in the RESULTS as a table e.g.

STATION	AMPLITUDE (mm)	DISTANCE from EPICENTRE (km)
Makasser		
Palangkaraya		
Tarakan		

3. Use the following chart, called a nomogram, to find the magnitude by placing a ruler or other straight edge between the distance (left hand column) and the amplitude (right hand column) for each of the three stations e.g. if the distance between station and epicentre was 300 km, and the amplitude was 100 mm, then the earthquake would have a magnitude of 6.0.

EXPERIMENT 26.1 continued

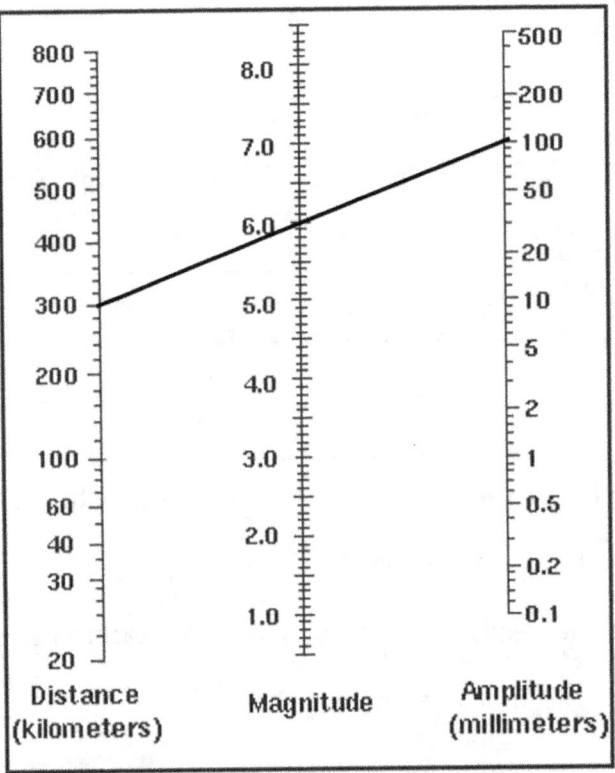

QUESTIONS:

1. In general, the amplitude should get smaller with distance from the epicentre. Why?

2. What other factors other than distance could reduce the amplitude of an earthquake wave?

3. Would such as earthquake trigger a tsunami? Why? Not all earthquakes near the sea produce tsunamis. Why?

CONCLUSION:

Give a report on the location and magnitude of this earthquake in Indonesia. Also suggest other problems which may occur as a result of this earthquake near the sea.

RESEARCH: (Optional)

Find out if an earthquake did occur in this region about September, 2018 and describe some of the problems associated with the earthquake as well as which the amount of destruction, casualties and any actions which may have reduced the damage. Give a Modified Mercali Scale value to this earthquake. Why was this earthquake especially bad?

EXPERIMENT 26.2

Time: one lesson

LOCATIONS of SOME MAJOR EARTHQUAKES

AIM: To plot the location of the epicentres for some major earthquakes to see if there is a pattern in their location.

MATERIALS: World map, table of major earthquakes and their locations

BACKGROUND:

There are several reasons why earthquakes occur. This activity is an exercise in plotting the latitude and longitude of some of the world's major earthquakes to see if there is a pattern in earthquake locations.

PROCEDURE:

1. Print or photocopy the following world map:

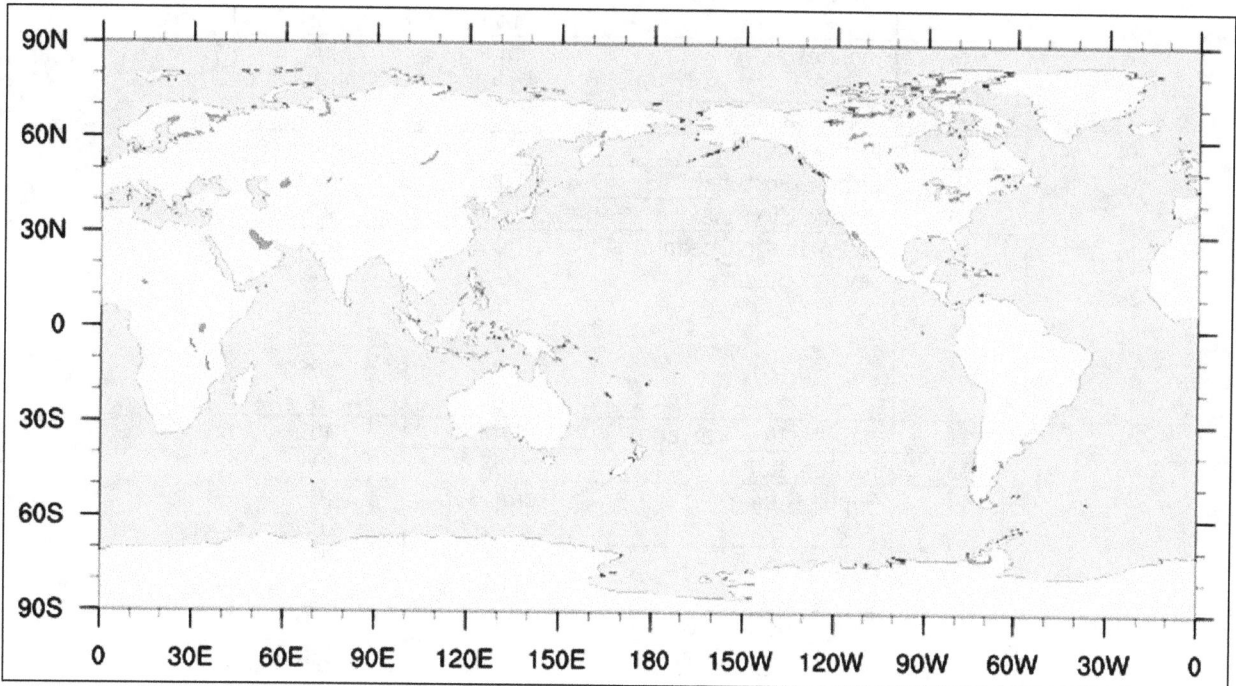

EXPERIMENT 26.2 continued
One lesson

2. Carefully plot the latitudes and longitudes for each earthquake given in the table below by using the numbers for these locations. Remember to plot latitude first and then the longitude. Some locations are regions and so may be shaded for the entire length between the positions given:

NUMBER	PLACE or REGION	LATITUDE	LONGITUDE
1	Aleutian Islands	from 165E to 170W	55N 52N
2	Anchorage, Alaska	152W	60N
3	Andes Mountains	From 75W to 70W	5N 55S
4	East Africa	from 40E to 35E	10N 20S
5	Himalaya Mountains	from 70E to 90E	30N 28N
6	Indonesia	from 95E to 130E	5N 10S
7	Kobe, Japan	135E	32N
8	Kurile Islands	150E	45N
9	Lisbon, Portugal	10W	40N
10	Mexico City, Mexico	105W	30N
11	Mid-Atlantic Ocean at several places:	0 40W 20W 0 60E 70E	80N 40N 0 50S 50S 0
12	Napier, New Zealand	178E	40S
13	Naples, Italy	15E	40N
14	Papua Nuigini	from 132E to 160E	0 20N
15	Phillipines	from 125E to 120E	5N 20N
16	Salaparuta, Sicily	12E	38N
17	San Francisco, USA	123W	35N
18	Taiwan	122E	22N
19	Tokyo, Japan	140E	35N
20	Turkey	30E	53N
21	Vanuatu	168E	15S

RESULTS:

Plot the location of each epicentre on the photocopy of the map using the numbers of each location. Students doing electronic reports will have to photograph or scan their completed map.

EXPERIMENT 26.2 continued

QUESTIONS:

1. How could a more accurate world pattern be obtained, especially in locations of isolated earthquakes?
2. Account for the epicentres in other locations which do not appear to have major earthquakes according to this data.

CONCLUSIONS:

1. Is there any pattern or logical groupings of epicentres?

2. Why do these epicentres exist in these locations?

3. How do these earthquake locations compare with those of volcanoes from Experiment 25.2?

4. From the map and other sources, list a few major cities which there could be valid predictions for:

    a. earthquakes
    b. tsunamis and
    c. volcanoes

5. Comment on each of the following statements:

    a. Earthquakes are caused by volcanoes
    b. Volcanoes are caused by earthquakes
    c. Earthquakes and volcanoes can have common sources

RESEARCH (Optional)

Use the Internet to find out about some other earthquakes e.g.

- Lisbon, Portugal in 1755
- New Madrid, USA in 1812
- San Francisco, USA in 1906
- Southern Chile in 1960
- Meckering, Australia in 1968
- Tanshan, China in 1976
- Loma Prieta , USA in 1989
- Newcastle, Australia in 1989
- Kobe, Japan in 1995
- Gujurat, India in 2001
- Haiti 2010
- Pueblo, Mexico in 2017

EXPERIMENT 26.3 Time: one lesson

PREDICTING EARTHQUAKES

AIM: To examine secondary data from an earthquake-prone area to see if future earthquakes can be predicted.

MATERIALS: Graph and data

BACKGROUND:

Earthquakes, volcanoes and tsunamis are the most dangerous Earth hazards to threaten large populations who live on or near the zones where these occur. Geophysicists, geochemists and oceanographers are able to use a range of remote-sensing devices to monitor hazardous zones. For example, regions of the world's oceans which can experience tsunamis such as the Indian Ocean and the Pacific Ocean have floating buoys and surrounding land-based monitoring stations which are linked by satellites. On active volcanoes, scientists can deploy temperature sensors, seismographs, tiltmeters and regularly test groundwater for height, temperature and additional dissolved gases. Earthquake zones with active faults are also monitored using seismographs and strain-meters and LASER geodometers can be positioned on either side of the fault to detect pressure within the fault and any slight slipping. It is very difficult to predict with certainty if an earthquake is going to occur. If the instruments show a dramatic rise or change in their data then all one can say is that there is a good probability that there may be an earthquake soon.

Faults which have a past history of sudden movement causing earthquakes can be monitored simply by measuring their slip or creep using LASER geodometers (which measure distance) at different points along the fault. The geodometer is set up at a specific point on the fault line and the distances to three different geographical points are measured. These distances are accurately triangulated to give the exact position of the fault. Over time, these points can be revisited and the movement of the fault measured.

PROCEDURE:

1. Look at the data table on the next page for the right-lateral movement (i.e. sideways motion to the right as the geodometer is facing) of the Calaveras Fault near Gilroy.

2. These values are Cumulative Slip over time. Cumulative slip is that movement which has occurred since all of the measurements were made i.e. total movement up to that year:

YEAR	SLIP (mm)	YEAR	SLIP (mm)
1970	10	1988	340
1972	25	1990	365
1974	40	1992	400
1976	60	1994	425
1978	80	1996	450
1980	120	1998	475
1982	175	2000	500
1984	225	2002	525
1986	285	2004	550
NOTE: values form 2006 to present go up uniformly at 25mm/year (Data and map from the USGS)			

3. Plot these values on a graph in the RESULTS as a line graph, noting the gradient or slope of each section.

EXPERIMENT 26.3 continued

One lesson

4. Extrapolate the graph to the present day and note any changes of gradient (i.e. slope of the line).

5. If there are any slight changes in the slope of the graph, these represent sudden movements along the fault which could cause an earthquake. Identify any such changes and research the Internet to find an earthquake of that year for California.

RESULTS:

Copy this graph template and complete this line graph from the provided data.

Locate the site of the measurements at Gilroy and the Calaveras Fault on the following map.

QUESTIONS:

1. What is a LASER geodometer? How does it work?

2. What is a cumulative right-lateral creep?

3. How will such measurements help with probabilities of future earthquakes?

EXPERIMENT 26.3 continued   Time: one lesson

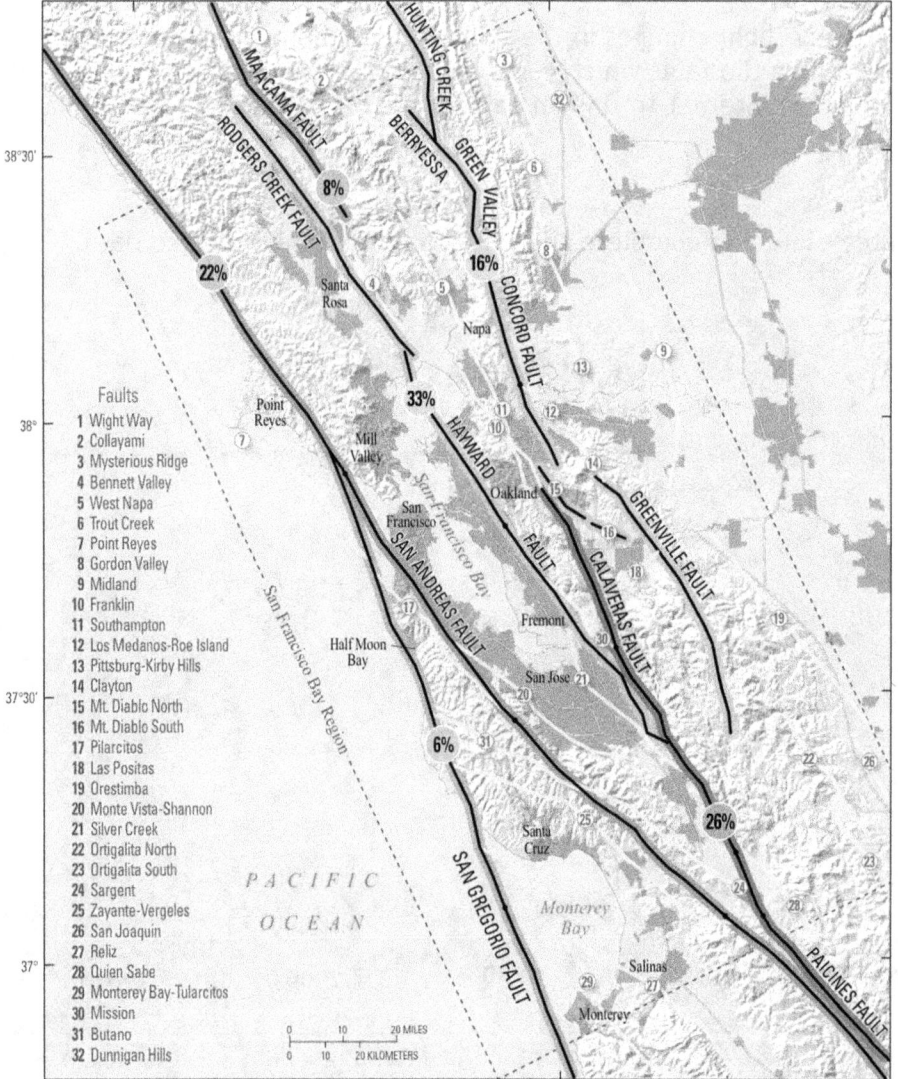

CONCLUSION:

What does the graph suggest about the possibility of an earthquake along the Calaveras Fault?

Comment on the use of such measurements in assisting with earthquake prediction. Comment on any correlation with Californian earthquakes since 1970 and their relationship to the graph. Also assess the difficulty which geophysicists would have in attempting to measure data in the attempt to predict earthquakes.

RESEARCH (Optional)

Use the Internet and reliable scientific sites to make an evaluation of the ability to predict earthquakes, volcanic eruptions and tsunamis.

Chapter 27: Wind, Rain and Fire

EXPERIMENT 27.1 Time: one or two lessons

RUNOFF COEFFICIENT and FLOODING

<u>AIM:</u> To examine some of the factors which determine runoff coefficient.

<u>MATERIALS:</u> Long tray or shallow trough with a hole or spout at its base at one end, shower or watering can rosette, measuring cylinder or beaker, several large wooden blocks, 100 ml measuring cylinder, watch/clock with second hand, sand, gravel, carpet square.

<u>BACKGROUND:</u>

The runoff coefficient (C) is a dimensionless coefficient relating the amount of runoff to the amount of precipitation received. It is a larger value for areas with low infiltration i.e. areas with less absorption into the ground, and with high surface flow such as pavements, steep gradients and open areas. It has a lower value for permeable surfaces with high infiltration, well vegetated areas, flat land and densely-packed built-up urban areas. It is important for flood control and for identifying for possible flood zone hazard areas. A high runoff coefficient (C) value may indicate flash flooding areas during storms as water moves fast overland on its way to a river channel or a valley floor. In very general and simplistic terms this means that the discharge of water across a surface (q) such as during flooding, depends upon the rainfall intensity (i) and the area it runs across (a) and the characteristics of the surface given by the Runoff Coefficient (C):

$$q \; \alpha \; C \, i \, a$$

where q is the Peak Discharge;
C is the Runoff Coefficient
i is the rainfall intensity; and
a is the area of the surface

The value of the ratio C depends upon the characteristics of the surface such as its gradient, its soil type or covering material and any obstructions placed upon it.

<u>PROCEDURE:</u>

PART A: Basic runoff

1. Set up a long, inclined tray which has a hole in one end at the base of its surface. The angle of inclination should be only slight (say 5 degrees). One end of the tray should be just below a source of water such as a tap with a watering can rosette (sprinkler) attached whilst the other end where the water will flow out, over a sink:

EXPERIMENT 27.1 continued

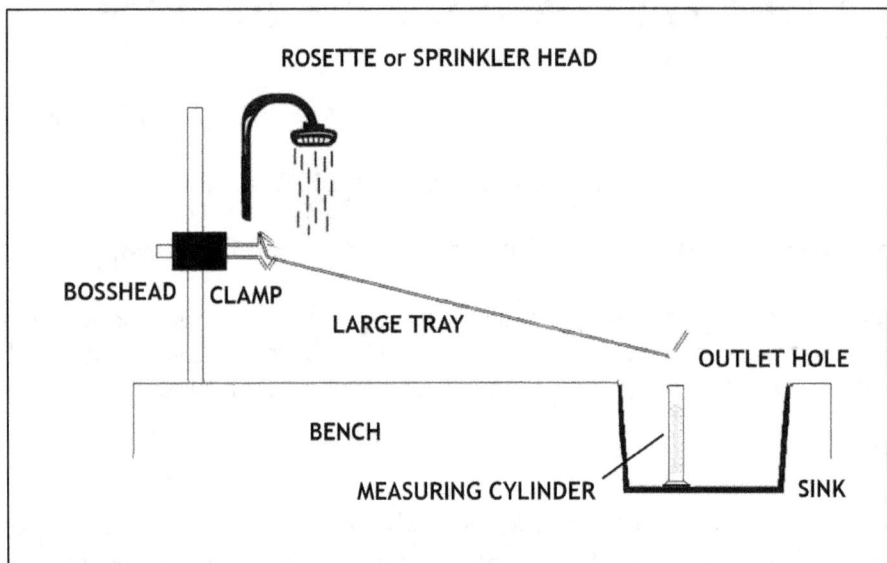

2. Adjust the water flow through the rosette so that it exactly matches the water flowing out into the sink. This should be relatively slow and might take some care and may need adjusting the tap or the angle of inclination of the tray. It is important that this angle be only small and that no water builds up as a pool in the tray.

3. With the water flow adjusted, representing a constant rainfall, measure the time that it takes to fill the measuring cylinder to the 100ml level. This represents the Peak Discharge. Record this value in the RESULTS.

4. The area (a) of the surface is fixed by the dimensions of the tray and is a constant. Whilst keeping the flow rate unchanged, change some of the other parameters at this gradient of the tray and measure the times to fill 100 ml in the measuring cylinder. Such changes could include:

    a. Placing a few large blocks of wood randomly on the surface to represent an urban area;
    b. Many large blocks of wood place together to represent an industrial estate;
    c. Remove the obstructions and line the bottom of the tray with a carpet square to represent grass;
    d. Replace the carpet with a thin layer of wet sand or gravel to represent different surface soil types;
    e. Remove everything from the tray and change the angle of inclination several times to represent different rainfall intensities.

It would be more accurate if each of the parameters started with a dry tank but a way of having the water flow kept at a constant rate as in the initial procedure would be needed. Also, small coloured beads could be dropped in at the water input to see if the water flows differently around the blocks.

5. Record all of the timings to fill the measuring cylinder to 100 ml level for each of these parameter changes in the results as a table.

EXPERIMENT 27.1 continued

RESULTS:

Record the readings as a table e.g.

PARAMETER	FLOW RATE/100 ml (seconds)
Empty tray (initial angle)	
Few wooden blocks	
Many wooded blocks	
Carpet no blocks	
Sand (wet, packed) or Gravel (wet, packed)	
Empty tray (angled at ^{0})	
Empty tray (angled at ^{0})	

Make any general comment about the observations made, especially about the flow rates around or over obstructions.

QUESTIONS:

1. Why is it important to have a constant water flow and not allowing it to pool in the tank?

2. Why should the carpet and the soil etc be wet before recoding the water output in the measuring cylinder?

3. What would be some improvements in the procedure to assist in better results?

4. This experiment attempts to keep the rainfall as a constant. How could the effects of increased rainfall be measured?

CONCLUSION:

Runoff Coefficient concerns the ratio of the water flowing off a surface compared to the rainfall falling upon it. Comment on the factors which will affect this ratio assuming that the rainfall is constant.

RESEARCH (Optional)

Use the internet to find out more about Runoff Coefficient, how it is measured and how it is used to reduce floods.

Chapter 28: A Changing Climate

EXPERIMENT 28.1 Time: one lesson

CORRELATION of CARBON DIOXIDE LEVELS and GLOBAL TEMPERATURE CHANGE

<u>AIM:</u> To see if there is a correlation, or match, between increased atmospheric carbon dioxide levels and temperature change.

<u>MATERIALS:</u> Computers or tablets with internet connection, calculators as required.

<u>BACKGROUND:</u>

Scientists usually have to deal with large pieces of mathematical data and often have to compare the data of one variable, such as global temperature increases to another, such as atmospheric carbon dioxide concentration. To see if there is a match, or correlation, between such data, they will use a mathematical statistical method called the Pearson Correlation Coefficient named after the statistician Karl Pearson (English:1857-1936). For two sets of values (e.g. set x and set y), the Pearson Correlation Coefficient (r) is given by:

$$r_{XY} = \frac{\sum (X_i - X)(Y_i - Y)}{\sum \sqrt{(X_i - X)^2} \sqrt{\sum (Y_i - Y)^2}}$$

Where X_i and Y_i are the individual values in each set;
X and Y are the means or averages of each set; and
Σ (the Greek letter sigma) is the sum or addition of these

Values of r will range between -1.0 and +1.0. If the calculated value is close to +1.0 then this means that the two sets of data are directly and closely matched i.e. if one increases then so does the other. If the value is close to -1.0 then there is a close but inverse correlation i.e. if one increases the other decreases.

This experiment uses averaged Temperature Anomalies as an indication of temperature change. These are the differences in temperatures (in Celsius degrees) compared to a calculated long-term average temperature.

<u>DATA:</u>

The following table of data was obtained from the National Oceanographic and Atmospheric Administration (NOAA). The values for the atmospheric carbon dioxide concentration (in parts per million ppm) were obtained from NOAA's Mauna Loa Observatory on the island of Hawai'i. Temperature anomaly values were obtained from various sites collected by the National Aeronautics and Space Administration (NASA).

EXPERIMENT 28.1 continued

YEAR	CARBON DIOXIDE CONC. (parts per million ppm)	TEMPERATURE ANOMALY (Celsius degrees)
1960	316.91	+0.02
1965	320.04	-0.04
1970	325.68	0.00
1975	331.11	+0.02
1980	338.75	+0.20
1985	346.12	+0.21
1990	354.39	+0.34
1995	354.39	+0.34
2000	369.55	+0.50
2005	379.80	+0.61
2010	389.90	+0.62
2015	400.83	+0.80

PROCEDURE:

1. Go to the following website which provides a Pearson Correlation Calculator:

    https://www.socscistatistics.com/tests/pearson/Default2.aspx

    This is used instead of one's own computer which has been programmed with the information for calculating Pearson Correlation.

2. COPY and PASTE the values for the CO2 concentrations and the temperature anomalies separately in the two columns provided for X and Y on the website and then press CALCULATE R.

3. This will give the value for the Pearson Correlation Coefficient as well as a graphical plot or Scattergram of how each value for the years match. A good linear plot will show a high correlation. A negative value for r will give an inverse plot.

RESULTS:

Make any general comment about using such statistical analysis and give the value for the Pearson Correlation Coefficient which has been calculated.

Also make a copy (electronically or otherwise) of the scatter plot obtained from the website after calculating the Correlation Coefficient. Join up the dots of the plot to make a graph and comment on this visual correlation.

EXPERIMENT 28.1 continued

QUESTIONS:

1. Why are temperature anomalies used instead of maximum or minimum temperatures?

2. Why is the data given as concentration of carbon dioxide in parts per million rather than as an emission value in tonnes?

3. Why are CO_2 concentrations measured on top of a volcano in Hawai'i? What problems are associated with measuring CO_2 concentrations?

4. Why is statistical analysis such as the Pearson Correlation Coefficient used?

5. Are there any problems or errors associated with this type of calculation (some Internet research needed)?

CONCLUSION:

1. What was the correlation coefficient between atmospheric carbon dioxide concentrations and global temperature?

2. What does this show about the relationship between these two variables over time? Discuss.

RESEARCH (Optional)

Use the internet to find out more global warming and carbon dioxide gas concentrations and emissions, especially the emissions of different countries. See:

https://ourworldindata.org/co2-and-other-greenhouse-gas-emissions

which gives interactive maps showing data for most countries. Compare the data for Australia, United States, China and some of the less-industrialised nations.

http://berkeleyearth.lbl.gov/city-list/

also gives the global warming data for major cities.

Also bookmark the following site which gives a daily measure of global carbon dioxide levels:

https://www.co2.earth/daily-co2

APPENDIX A: Risk Assessment of Practical Work and Excursions

It has become mandatory that a risk assessment is made before carrying out any laboratory practical activity or excursion. These must consider:

1. the nature of the activity;

2. equipment, chemicals, living things and other items to be used;

3. the nature of the environment, especially if venturing outside for an excursion, even in the local area; and

4. the student.

Once all of the factors and options have been considered, a Risk Assessment Form should be completed and kept in a central records section (often kept by a Laboratory Manager). It is also a good idea for a copy to be kept (electronically and print) by the teacher along with the set of notes for the experiments and activities (perhaps attached to the teacher's copy of this book). Once completed, it should be revised and modified as required each time the activity is performed. This need not be an onerous task after it has been done for the first time. Thereafter it simply requires a quick review of the previous form and modifications if any of the variables (e.g. different student type, different location, new equipment, different teacher etc.) have changed.

Safety is paramount for any teaching activity and the students must be made aware of any potential hazard in the activity and take adequate precautions. Some suggestions for laboratory safety and safety during field excursions have been given at the beginning of this book. It is always useful to have these suggestions displayed prominently in the laboratory. In the associated textbook ADVENTURES IN EARTH and ENVIRONMENTAL SCIENCE, each chapter contains a PRACTICAL TIPS at its conclusion. These have been based on over forty years of safe laboratory and field practices as a field researcher well as many years travelling into some of the most hostile and remote parts of the worlds, such as the Antarctic Peninsula, Amazon Basin, north African deserts and alpine regions in New Zealand, Europe and the Andes. This section also outlines, as does various, parts of the body of the textbook, how science is one of the most interesting endeavor of Humankind.

Institution laboratory managers should have on file a detailed indexed catalog of the risk or hazard potential of all chemicals, electrical equipment, biological specimens (including live organisms) and other items in use within the laboratories.

A simple Risk Assessment Form is given on the next page:

Earth and Environmental Science
Risk Assessment Form

DATE:	CLASS:	LOCATION:

ACTIVITY:	TYPE (Pract./Excursion/visit):

RISK LEVEL : (High/Medium/Low)	Students:	Teacher:	Other(who/what):

OUTLINE of ACTIONS	**RISK**	**PRECAUTION/ACTIONS**

DISPOSAL of WASTES:

OTHER COMMENTS:

A more detailed Risk Assessment Form can be found at:

https://www.aisnsw.edu.au/workplace-health-and-safety/Documents/Appendix_A_Science_and_Technology_Risk_Assessment_Template_Rev.1.docx

Some useful references for risk assessment can be found at:

https://education.qld.gov.au/sitesearch/Pages/results.aspx#k=risk%20assessment

https://education.qld.gov.au/initiatives-and-strategies/health-and-wellbeing/workplaces/safety/managing/risk-management

https://www.riskassess.com.au/info/routine_safety_procedures

https://smah.uow.edu.au/content/groups/public/@web/@sci/@chem/documents/doc/uow016874.pdf

https://www.riskassess.com.au/docs/RABrochureAU.pdf

http://www.nswtitration.com/files/school_risk_assess.pdf

https://assist.asta.edu.au/

https://assist.asta.edu.au/search?expert=&field_curriculum_year=All&field_publication_date=All&keywords=risk+assessment&field_tax_australian_curriculum_parent_parent_parent_tid=&laboratory_technicians=All&area=&field_voting_user_rating=&field_voting_average_rating_1=&field_voting_user_rating_1=&rating_point=All&year_level=&sort_by=created

https://education.nsw.gov.au/teaching-and-learning/curriculum/key-learning-areas/science/safety

https://www.education.vic.gov.au/school/principals/spag/governance/Pages/riskinplanning.aspx

http://ascip.org/wp-content/uploads/2016/06/ASCIP-Risk-Management-Primer-for-School-Districts-SIXTH-DRAFT-2016-06-20.pdf

https://www.rospa.com/rospaweb/docs/advice-services/school-college-safety/managing-safety-schools-colleges.pdf

https://www.leeds.ac.uk/secretariat/documents/risk_management_guidance.pdf

https://www.teachers.org.uk/files/safety-in-practical-lessons_0.doc

APPENDIX B: Excursion Permission Note

Some institutions often require an Excursion Permission Note from students who are under legal age who are leaving the institution for a field, industrial/scientific visit or other outside activity. These DO NOT absolve the teacher and guides of any moral and legal responsibility but merely provides parents/guardians with the necessary information as to what their child will be doing and where. It also can provide a list of safety and comfort items which will be needed from home for the activity in the hope that such reasonable items will be supplied.

Some examples of field or other outside activities for this program include:

- Rock Quarries
- Mining areas (special permission and rules from the company)
- Freshwater stream ecology
- Marine Rock Platform ecology
- Rainforest/Dry Forest/Grasslands ecology
- Mining/mineral Museums
- Environmental Stations (special rules and guides apply)
- Museums of Natural History
- Scientific Research Organisations (special rules and guides apply)
- National Parks (special local rules and guides necessary)
- Government Agencies (special rules and guides apply)

A typical activity Permission Note is given on the next page and may be copied for the students to take home:

< insert institution letter head>

EARTH and ENVIRONMENTAL SCIENCE EXCURSION

DATE: <insert date and times>   **CLASS:** < insert class/group>
DESTINATION: <insert location/name of place or organisation etc.>

This excursion is an integral part of the semester's Programme. All students will be required to complete an assignment, associated with the excursion, which will contribute towards the assessment of the subject.

ARRANGEMENTS: < insert details of transportation, time of departure and return etc.>

The bus will leave at   < time>   and returns to the school at   < time>   approx.
(Bus company can be contacted at: <name and contact number>   )

Students are to meet at: < meeting place and time>
but will not to enter the bus until directed to do so. The rules of the School apply at **all times**

Staff going on the excursion will be: < name(s) of staff and contact numbers >

SPECIAL REQUIREMENTS: < special requirements such as dress, personal items such as cameras, mobile phones, writing material, safety precautions etc. Also any special rules of behaviour and group actions if lost etc.>

Yours faithfully,
< name of person in authority>

Head of Earth and Environmental Science

CONSENT FORM: Please return to supervising teacher by   < insert time/date>

I have read the above information and agree for...... **<insert student name>** ....to go on the excursion.

Special information concerning my daughter's/son's welfare that the supervising staff should know is as follows:

<insert special requirements such as allergies, dietary requirements, medical conditions etc.>

Phone contact of parent/guardian   Home:..........................Work..............................

Signature:..(Parent/Guardian)

Date ……………………………….

Books by the Author

ADVENTURES IN EARTH AND ENVIRONMENTAL SCIENCE - BOOK 1. This is the first book of two which looks at the Earth, its matter, energy relationships and its life and how all interact together as a whole. The Earth is seen as a closed system contain the Earth's materials and living things but allowing a necessary flow of energy into and out of the planet. The atmosphere, hydrosphere, geosphere and biosphere of Earth are all examined in detail and lavishly illustrated with over five hundred photographs and diagrams. There are also several links to videos made by the author during his own adventures in studying the Earth. Each chapter is concluded with a Summary, Practical Tips, ten Multichoice Questions and ten longer Review and Discussion Questions. There is also an accompanying PRACTICAL MANUAL with a large number of experiments and data analysis activities which can be performed by students using basic available equipment in support of the textbook. This manual also teaches students how to investigate and write research reports for submission.

ADVENTURES IN EARTH AND ENVIRONMENTAL SCIENCE - BOOK 2. This is the second book with this title and, after a short revision chapter, looks at how Humankind lives on planet Earth. Renewable and non-renewable resources are described and how their use has impacted on the world's ecosystems, on land, in the sea and in the atmosphere. This is also discussed with an emphasis on the problems of future energy needs, global warming and social consequences of these events. As well as problems caused by Humankind, the natural hazards of the Earth have also been described with the view that many or the world's populations live in regions that can be very dangerous at times. The contents of this book are also supported with many photos, illustrations and videos. Each chapter is concluded with a Summary, Practical Tips, ten Multichoice Questions and ten longer Review and Discussion Questions. Book 2 also has an accompanying PRACTICAL MANUAL with a large number of experiments and data analysis activities which can be performed by students using basic available equipment in support of the textbook.

ADVENTURES IN EARTH AND ENVIRONMENTAL SCIENCE is the composite book containing all of the content of Books 1 and Books 2. It has been written as a utilitarian reference book for the classroom, library or home study.

ADVENTURES IN EARTH AND ENVIRONMENTAL SCIENCE - TEACHERS' GUIDE has been designed to assist teachers with the use of these books and the teaching of this subject. It gives advice on lesson p[reparation, the teaching of the practical work and answers to the questions contained in the books.

Adventures in Earth and Environmental Sciences Books 1, 2 , the composite book and the Teachers Guide are all available in electronic (Kindle) format which can be viewed using any electronic device having the free Kindle App. They are also available in PRINT editions from Felix Publishing at:

(info@felixpublishing.com)

ADVENTURES IN EARTH SCIENCE

ADVENTURES in EARTH SCIENCE is not just an in-depth and traditional Earth Science textbook but a series of adventures across seven continents and beyond in the sciences of astronomy, geology, meteorology and oceanography. It has been written with over forty years of experience in studying, researching and teaching earth science. Whilst it has been designed for senior high school and junior university or college, it is written in an easy style and well-illustrated so that anyone with an interest in this topic would find it an interesting and valuable resource.

The latest scientific information has been given in the text including chapters on climate change and the future use of fuels and energy. The book contains over 700 pages, 1200 photographs and illustrations mostly taken by the author. It also includes 32 video links taken by the author to explain various skills as well as excursions to many exotic places in support of the text.

Each chapter is concluded with a summary of the main points, multichoice and extended review questions and a section of helpful hints which offers practical suggestions in such matters as collecting specimens, field work and survival, astronomical observation and general observation and techniques. Answers to the multichoice questions are given at the end of the book as is a 38 page key index and glossary section which gives the main page references as well as definitions of important terms fold as bold type in the text.

ADVENTURES IN EARTH SCIENCE - STUDENT PRACTICAL MANUAL

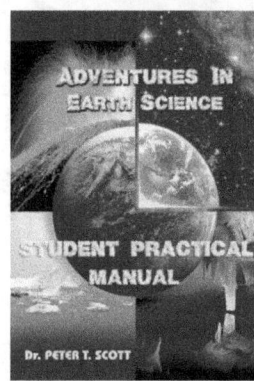

This student includes sixty formal student experiments in support of the textbook ADVENTURES in EARTH SCIENCE. Each experiment is set out in the traditional format and advice is given so that students may write formal reports on their findings. These experiments have been used successfully in a career spanning over forty years with the last twenty-one being devoted to Earth Science.

These experiments are not mandatory and the teacher may change the order of the experiments, delete some or add others depending upon the curriculum needs and the equipment and time available.

164 pages 67 diagrams

ADVENTURES IN EARTH SCIENCE - TEACHER'S GUIDE

This is the teacher's guide for the textbook ADVENTURES in EARTH SCIENCE and its companion STUDENT PRACTICAL MANUAL. It contains educational and scientific advice from over forty years of practical teaching experience and field work for the teaching of Earth Science.

The Introduction contains useful information for the teaching and administration of this subject designed for the teacher who may have had minimal experience in these matters.

Subsequent chapters which relate to those of the textbook and Practical Manual provides useful information including advice in setting up and assisting students with the practical work, sources of materials which may be required, additional activities with many video link references and Internet resources. The answers to the Multichoice Questions of the textbook are given as are suggestions for the longer Review Questions. The references given in the textbook are supplemented by many additional references.

275 pages   41 diagrams   62 video links

The contents of this book have also been rearranged into the **ADVENTURES IN EARTH SCIENCE** SERIES of smaller individual electronic books also available through amazons.com. They will also be made available soon in smaller A5 print editions through Felix Publishing (info@felicpublishing.com).

These include:

 Exploration Science: Field Geology and Mapping

 Fossils- Life in the Rocks -

 Riches from the Earth: Minerals, Energy and Mining

 A Dangerous Planet: Volcanoes and Earthquakes

 Changing the Surface: Erosion and Landscapes

 Through Sea and Sky: Oceanography and Meteorology

 Rocks - Building the Earth

 Beyond Planet Earth: Introduction to Astronomy

All of these books are available in electronic format which can be purchased at amazon.com for any PC or tablet in Kindle format which can be read on any device using the free Kindle App.

Contact Felix Publishing for details at info@felicpublishing.com

www.ingramcontent.com/pod-product-compliance
Lightning Source LLC
Chambersburg PA
CBHW050713090526
44587CB00019B/3365